Narrative Thinking and Storytelling for Problem Solving in Science Education

John Thomas Riley
The Big Moon Dig, USA

Luisa dall'Acqua
Lyceum TCO Italy, Italy

A volume in the Advances in Early Childhood and K-12 Education (AECKE) Book Series

IGI Global
DISSEMINATOR OF KNOWLEDGE

Published in the United States of America by
IGI Global
Information Science Reference (an imprint of IGI Global)
701 E. Chocolate Avenue
Hershey PA, USA 17033
Tel: 717-533-8845
Fax: 717-533-8661
E-mail: cust@igi-global.com
Web site: http://www.igi-global.com

Library of Congress Cataloging-in-Publication Data

Names: Riley, John Thomas, 1945- author. | Dall'Acqua, Luisa, 1962-, author.
Title: Narrative thinking and storytelling for problem solving in science education / By John Thomas Riley and Luisa dall'Acqua.
Description: Hershey : Information Science Reference, [2019]
Identifiers: LCCN 2018053031| ISBN 9781522584018 (hardcover) | ISBN 9781522584025 (ebook)
Subjects: LCSH: Science--Study and teaching (Higher) | Science--Methodology. | Storytelling. | Narration (Rhetoric)
Classification: LCC Q181 .R563 2019 | DDC 507.1/1--dc23 LC record available at https://lccn.loc.gov/2018053031

This book is published in the IGI Global book series Advances in Early Childhood and K-12 Education (AECKE) (ISSN: 2329-5929; eISSN: 2329-5937)

British Cataloguing in Publication Data
A Cataloguing in Publication record for this book is available from the British Library.

Advances in Early Childhood and K-12 Education (AECKE) Book Series

ISSN:2329-5929
EISSN:2329-5937

Editor-in-Chief: Jared Keengwe, University of North Dakota, USA

MISSION

Early childhood and K-12 education is always evolving as new methods and tools are developed through which to shape the minds of today's youth. Globally, educational approaches vary allowing for new discussions on the best methods to not only educate, but also measure and analyze the learning process as well as an individual's intellectual development. New research in these fields is necessary to improve the current state of education and ensure that future generations are presented with quality learning opportunities.

The **Advances in Early Childhood and K-12 Education (AECKE)** series aims to present the latest research on trends, pedagogies, tools, and methodologies regarding all facets of early childhood and K-12 education.

COVERAGE

- Common Core State Standards
- Diverse Learners
- Standardized Testing
- Early Childhood Education
- Poverty and Education
- Special Education
- K-12 Education
- Reading and Writing
- Head Start and Pre-K Programs
- Bullying in the Classroom

IGI Global is currently accepting manuscripts for publication within this series. To submit a proposal for a volume in this series, please contact our Acquisition Editors at Acquisitions@igi-global.com or visit: http://www.igi-global.com/publish/.

The Advances in Early Childhood and K-12 Education (AECKE) Book Series (ISSN 2329-5929) is published by IGI Global, 701 E. Chocolate Avenue, Hershey, PA 17033-1240, USA, www.igi-global.com. This series is composed of titles available for purchase individually; each title is edited to be contextually exclusive from any other title within the series. For pricing and ordering information please visit http://www.igi-global.com/book-series/advances-early-childhood-education/76699. Postmaster: Send all address changes to above address.

Titles in this Series

For a list of additional titles in this series, please visit:
https://www.igi-global.com/book-series/advances-early-childhood-education/76699

Critical Essays on the New Moral Imperative for Supporting Marginalized Students in PK-20 Education
Rebecca J. Blankenship (Florida Agricultural and Mechanical University, USA)
Information Science Reference • ©2019 • 268pp • H/C (ISBN: 9781522577874) • US $175.00

Cyberbullying and the Critical Importance of Educational Resources for Prevention and Intervention
Gilberto Marzano (Rezekne Academy of Technologies, Latvia) and Joanna Lizut (Janusz Korczak Pedagogical University in Warsaw, Poland)
Information Science Reference • ©2019 • 230pp • H/C (ISBN: 9781522580768) • US $175.00

K-12 STEM Education in Urban Learning Environments
Jillian L. Wendt (University of the District of Columbia, USA) and Danielle L. Apugo (University of the District of Columbia, USA)
Information Science Reference • ©2019 • 309pp • H/C (ISBN: 9781522578147) • US $190.00

Cases on Kyosei Practice in Music Education
Richard Keith Gordon (Seisa University, Japan & California State University, USA) and Taichi Akutsu (Shujitsu University, Japan & Seisa University, Japan)
Information Science Reference • ©2019 • 296pp • H/C (ISBN: 9781522580423) • US $175.00

Cultivating a Culture of Nonviolence in Early Childhood Development Centers and Schools
Simon George Taukeni (University of Namibia, Namibia)
Information Science Reference • ©2019 • 322pp • H/C (ISBN: 9781522574767) • US $185.00

Advanced Strategies and Models for Integrating RTI in Secondary Schools
Pam L. Epler (Grand Canyon University, USA)
Information Science Reference • ©2019 • 356pp • H/C (ISBN: 9781522583226) • US $185.00

For an entire list of titles in this series, please visit:
https://www.igi-global.com/book-series/advances-early-childhood-education/76699

IGI Global
DISSEMINATOR OF KNOWLEDGE

701 East Chocolate Avenue, Hershey, PA 17033, USA
Tel: 717-533-8845 x100 • Fax: 717-533-8661
E-Mail: cust@igi-global.com • www.igi-global.com

Table of Contents

Preface vii

Acknowledgment xxxiv

Section 1
The Big Moon Dig Stories

Chapter 1
Consideration of a Grass-Roots Space Program: A Didactic Introduction 1

Chapter 2
The Humanity of an Artificial Intelligence 36

Chapter 3
AI and the Movies: A Storytelling About Communication 63

Section 2
Iron Seas Stories

Chapter 4
Technical Measurement, Control, and Forecasting of Climate Change 99

Chapter 5
The Greenland Ice: A Story of Changing Climate 123

Chapter 6
Ocean Acidification: A Problem for Policymakers 146

Section 3
The Herd Stories

Chapter 7
In Action on Desertification: A Didactic Introduction 171

Chapter 8
Consideration of Virtual Reality as Training: A Didactic Introduction............. 199

Chapter 9
AI and People: A Storytelling About Interactions 224

Chapter 10
AIs Training People: A Storytelling About Building Life Teams 245

About the Authors 269

Index 270

Preface

The 21st century has no shortage of historic problems; global warming just for a start. This places a major burden on society's young people and they need a clear vision of their place in a workable future.

Starting in the fall of 2018, a series of major studies on global warming became available. In October 2018 came the release of United Nations Intergovernmental Panel on Climate Change Special Report 1.5 (IPBES, 2019), and hard on its heels, the United States Fourth National Climate Assessment then not to be out done, the British report "This is a Crisis" (Laybourn et al., 2019). The UN report has the most complete science; the American report has what Congress needs; the British report is the best written and consequently has the best quotes.

These three reports all lay out action plans necessary (1) to address many of the great problems that are now certain to happen (remember Puerto Rico), (2) to prevent some of the worst effects of climate change (for example through sequestering carbon), and finally (3) to progress toward human societies for a sustainable Earth. Enabling these action plans will require an effort comparable to that required to win World War II. Failing to organize such an effort could result in unacceptable damage to our beloved institutions in addition to a great loss of jobs and treasure. Fortunately, the basis for action has already been studied (Hawken, 2017).

Above all, this effort will require an enormous number of people in effective action, and this burden will fall largely on today's young people. How is society preparing our young people to take on these challenges?

Currently in the mass media (TV, movies, and video games) society provides extensive lessons, costing hundreds of millions of dollars to produce, in addressing a zombie apocalypse or a robot insurrection. What is more, the stories show a particular interest in how these problems would be handled by comic book heroes.

It is clear that zombies in the streets and robots run amuck are not the great problems our young people will have to face. Furthermore, comic book solutions simply will not work in real life.

Society is now teaching our youth to fight the wrong wars using the wrong strategies. The result could be truly deadly.

Making an effort on this problem starts with the process of generating better instructional material as a grassroots effort. There are plenty of historic problems and a fair number of obscure but promising approaches that warrant testing but do not currently attract the level of attention needed to find and secure the necessary resources for a proper test.

This then is a great story opportunity. Attractive characters putting out a heroic effort and showing the personal devotion needed to prove out possible world-changing solutions makes for great stories.

WHAT IS STORYTELLING?

Storytelling is one of the most widespread, powerful, and multifaceted forms of communication. It concerns actions and events, related to real or invented facts, stimulating curiosity and imagination, and reflections.

The history of storytelling has developed parallel to the history of human culture. Its expressions since the beginning of civilization, responding to the fundamental need to share experiences, fix social and religious values, provide entertainment, explain phenomena and events natural and historical, but also: "educate", and "transmit".

From the rupestrian recordings stories of heroes recited by the Aedes in Greece with the accompaniment of music to the poems to the cosmogonies, storytelling proves to be the privileged communicative form both for the transmission of tradition and cultural identity of a population, building and sharing a system of values, symbols, and ideas (Bruner, 1990).

Stories can engage attention, influence our beliefs or actions, and provide a "partial suspension of the rules of the real" that helps people safely explore the future. It can mean an approach in which groups of people participate in gathering and working with raw stories of personal experience in order to make sense of complex situations for better decision making.

Storytelling usually is characterized by the presence of a narrating voice, a point of view, and implicit or explicit logical constraints (Aylett, 2006; Schank, 1990) defines storytelling below:

- A plot (or "fabula"): an overall configuration of the sequence as a whole, organized around a time, with a closure, a conclusion, a resolution. The narrative activity does not consist simply in approaching one episode to another, but also in the construction of significant totalities starting from distinct events (Ricoeur, 1981).
- A knowledge representation: a way of expressing a series of events in a connected way from both a causal and a temporal point of view (Herman, 2003).
- An organizational principle: a way of organizing perceptions, thoughts, memory, and actions (Wertsch, 1998). The ability to narrate, as well as the corresponding ability to follow a story, therefore requires that the student to outline a configuration starting from a simple succession (Polkinghorne, 1988).
- A cognitive process: a way of integrating significant cognitive mechanisms, such as reasoning and visual thought (Luckin et al., 2001; Scalise Sugiyama, 2001). It conveys meaningful and strong messages, structured according to a cause-effect logic.
- A categorization: a conceptualization, according to a bottom-up path, that is to say from the particular to the general, tending to the progressive abstraction (Bruner, 1990).

A narrative schema can be logically different, such as:

- A classic scheme in which the narrative begins at the center of an action, to then explain the beginning of the story and prepare its conclusion
- The classic scheme centered on the figure of a hero who abandons his home to embark on a journey to unknown places
- A pathway of the distribution of tension until a peak is reached and the subsequent descent (typical of the television series)
- A circular scheme in which narration containing a central message, interacting with others used as an argumentative and demonstrative text
- A narrative structure in which the discourse develops on two opposing levels that continually intertwine and represent the one "as things are" and the other "how things should be "
- A discursive structure in which different strands of thought converge to form a single idea. It can be used to show how an idea is the result of multiple paths leading to it

- The narrative begins with an apparently predictable plot that stops abruptly to give rise to a new beginning
- A discursive structure to organize multiple stories that move around the same central concept

The language of narration is acquired through exercise and use, so the child learns what, how, where, and to whom to speak. When the subject is able to designate things by name, his linguistic interest focuses on human action and its results and on the interaction between individuals.

How to Use Storytelling in Teaching

The term "storytelling" originated in the United States as "Storytelling Management," around the end of the twentieth century, as a strategy of Business Management and Marketing. It has also been widely used in the political sphere. In all its forms, a story is a dialectic between described events and the expectation of their development, an invitation to formulate problems (problem posing), but really not a concrete lesson on how to solve them" (Bruner, 2003, p. 20).

Specifically, in the educational field, it means a set of narrative and rhetorical techniques, a natural tool which can connect the knowledge building with specific learning activities, suitable for any learning object. It can support learning contexts creation (Aylett, 2006).

Specifically, it can be used for:

- Creating and using stories, suitable for transmitting incisive knowledge contents (McEwan & Egan, 1995).
- Motivating people to learn (Jackson, 1995).
- Organizing context (James & Minnis, 2004).
- Researching and developing procedures to extract and interpret data from narratives (Lieblich et al., 1998).
- Building maps of the world the student experiences so the student can make decisions about how to act.
- Making decisions about what to believe in, what the student can take to heart.
- Transfer knowledge and information.
- Playfully simulating possible outcomes before the student commits to a course of action.

- Condensing experience into packages that re-expand in the minds of listeners.
- Communicating ideas, experiences, knowledge.
- Building interpretative meanings of reality, a correct context to the environment in which the topic is placed.
- Providing a didactic output or a practical play activity.
- Offering an opportunity to valorize each person's experiences, and exchange experiences.
- Finding a valuable assumption (the "moral" of the story; Aylett, 2006).

A story generates other stories, according to the mechanism of intertextuality, favoring the collaborative exchange of knowledge, dialogic confrontation, the critical spirit and the search for new interpretations and points of view on a problem and/or theme. Some of the main factors of success are:

- The strongly gratifying character of a narrative approach.
- A simpler access to complex concepts, a method already used by the ancient mythology, and philosophers such as Plato, who made extensive use of myths (stories) in his dialogues.
- Generating hermeneutic – interpretative processes and meaningful conceptual correlations.
- The ease of memorization of the story on the cognitive level.
- The ability of the narrative approach to favor networked knowledge and creativity.

The place of hyper-textuality as an educational technology is derived from those positions that described it as a form of organization of knowledge that mirrors its mental structure. Nonaka defines the "hypertextual organization" as a non-linear development potential, based on incremental free associations of communication and learning. He states that every organizational learning is given by the combination of people's knowledge and the specific way of integrating them, and this combination expresses its own cognitive "personality" (Nonaka & Takeuchi, 1995).

Narrative teaching takes advantage of the story, of the construction of short sketches, of role-playing games, provided they are correctly referred to as "the scientific object" that one wants to study. It does not exercise on results, but on processes, the dynamism of interest, involvement, sharing. The educational path becomes an itinerary of change, and the evaluation tools become qualitative, narrative and descriptive, and comprehensive (borrowed from the psychological and psychosocial disciplines).

Stories can be seen as paths that link together, in various ways (symbolic, analogical, causal, etc.), different points of the cultural continuum of a civilization, creating an order in it, building a possible sense in the complex, contradictory and multifaceted plot of events, memories, values, symbols, etc. which constitute the tradition of an oral culture.

The codes of representation of reality for teaching purposes, according to Damiano, can be described through four distinct categories, called "mediators of reality". By "mediation" the author means the facilitating action in the transition from the object of knowledge to the knowing subject.

The teacher, in the process of teaching communication, employs various forms of mediation. The different teaching methods use the mediators differently.

The teacher can hypothesize that the variety of mediators used is an important quality element if the teacher keep in mind that the learning styles of pupils can lead them to favor different forms of representation of reality, also in relation to the different recourse they make to their plurality "intelligence".

According to Damiano (2010), this mediation process is classified in: active, iconic, analogical, symbolic mediators. They have the task of allowing this "mediation" between reality and its representation: they act as a link between subject and object in the production of knowledge replacing reality.

- **Active Mediators**: They are the closest to reality, to the scripts of perceptual representation. They make use of experience and have a physical-perceptive consistency and emotional-affective density (e.g., direct experience and exploratory excursions).
- **Iconic Mediators:** They represent the characteristics of reality through graphics and spatial language. They replace their images with objects.
- **Analogical Mediators:** They refer to the learning possibilities inherent in the game and in the simulation (from dramatization to role-playing and to simulation games of various complexity). They allow the student to experiment with complex concepts or relationships in a protected context since it is fiction. They are the least used and the least known, but involving and incisive, both on the level of emotions and that of knowledge.
- **Symbolic Mediators:** They are the closest to conceptualization. They use numbers, letters, concepts, codes of representation, which are more universally used, where the use of words and abstract symbols is

dominant. Charts, maps, and schemes are on the border between iconic and symbolic mediators.

These mediators are tools through which the teacher focuses on the "intermediate zone" between teaching and learning, which is useful both to the student and to the teacher: for the student it means having the opportunity to personally and actively elaborate the known, while the teacher does not run the risk of imposing his own teaching and learning style to the students.

The teacher himself becomes an epistemologist, because in fact, in his teaching practices he expresses a theory of knowledge in the continuous effort of constructing meaning (Damiano, 2007) within situations where uncertainty prevails, the indeterminacy, sometimes vagueness and conflict. Even more so in proceeding in stages towards a personalized education, which leaves ample managerial margins also for students.

In this direction the characteristics of the trainer are more and more those of the reflective practitioner (Shön, 1987; Mezirow, 1991; Altet, 2009), the design modeling proposed by Lesh and Doerr and Gero's approach (Lesh & Doerr, 2003), the relationship between research and practice it becomes the one described in the New Didactic Research by E. Damiano (2013). In this context, teacher practice also plays a central role as an epistemological resource (Rossi, 2010).

DIGITAL STORYTELLING

Digital Storytelling or narration realized with digital tools (Web Apps, Webware) consists in organizing contents selected from the web in a coherent system, supported by a narrative structure, in order to obtain a story consisting of multiple elements of various formats (video, audio, images, texts, maps, etc.).

Digital storytelling employs digital technologies to create hypermedia narratives. There are various types of storytelling available depending on the media and the models that the producer decides to adopt. This choice sets the criteria on the basis on which story is organized.

Main advantages to using digital storytelling are:

- A very simple access to everyone thanks to web applications, which allow students to create stories using resources found on the web and to enrich them.

- A richness and variety of stimuli and meanings, deriving from the high information density and an intelligent user interface as a mash of codes, formats, events, characters, and information, which interact with each other through multiple paths and different analogical relationships.
- Being a form of narration particularly suited to communication forms such as those of journalism, politics, marketing, autobiography, and teaching.

Some common types of digital storytelling are:

- Storytelling as Timeline, a form of a narration of chronologically ordered events.
- Story mapping a form of storytelling that uses geographic maps or images to insert in them a series of links to web resources related to a specific issue in order to obtain a navigable route.
- Transmedia storytelling to create a story organizing in an environment, modeled on the News magazine or on the Presentation, resources available on the web of various formats (images, videos, animations, texts, sounds, music, news, etc.), related to a given wind or theme or problem or character, in order to obtain a multimedia and hypertextual story. This form of storytelling is particularly used in the journalistic field.
- Visual Storytelling the story is told through the use of images.
- In video storytelling the story is made through the ability to manipulate videos by inserting text, links, annotations, images, questions, etc. The result is interactive videos that can contain multimedia elements within them.

The possibilities of using an image are varied. Images can be arranged in series as in a presentation or slideshow and accompanied by links, texts, or the recorded voice of a narrator. They may be accompanied by links to multimedia resources and/or the recorded voice of a narrator. The possibility is there to make the image interactive so that, by clicking on it, the student can open resources on the web, or to describe an experience through the collection of images, creating albums or picture boards accompanied by short captions.

THE SHORT STORY

A "short story" is a brief fictional prose narrative that is shorter than a novel and that usually deals with only a few characters. It is usually concerned with a single effect conveyed in only one or a few significant episodes or scenes.

The character is disclosed in action and dramatic encounter but is seldom fully developed. Typical ancient short stories are jests, anecdotes, studied digressions, short allegorical romances, moralizing fairy tales, short myths, and abbreviated historical legends.

The short story is usually concerned with a single effect conveyed in only one or a few significant episodes or scenes. The form encourages the economy of setting, concise narrative, and the omission of a complex plot; character is disclosed in action and dramatic encounter but is seldom fully developed. Despite its relatively limited scope, though, a short story is often judged by its ability to provide a "complete" or satisfying treatment of its characters and subject.

Before the 19th century, the short story was not generally regarded as a distinct literary form. Jests, anecdotes, studied digressions, short allegorical romances, moralizing fairy tales, short myths, and abbreviated historical legends didn't constitute a short story. The prevalence in the 19th century of two words, "sketch" and "tale," affords one way of looking at the genre.

Basically, the tale is a manifestation of a culture's un aging desire to name and conceptualize its place in the cosmos. It provides a culture's narrative framework for such things as its vision of itself and its homeland or for expressing its conception of its ancestors and its gods. Usually filled with cryptic and uniquely deployed motifs, personages, and symbols, tales are frequently fully understood only by members of the particular culture to which they belong.

The sketch, by contrast, is intercultural, depicting some phenomenon of one culture for the benefit or pleasure of a second culture. Factual and journalistic, in essence, the sketch is generally more analytic or descriptive and less narrative or dramatic than the tale. Moreover, the sketch by nature is suggestive, incomplete; the tale is often hyperbolic, overstated.

As a genre, the short story received relatively little critical attention through the middle of the 20th century. Innovative and commanding writers emerged in places that had previously exerted little influence on the genre. Luigi Pirandello, Franz Kafka, Jorge Luis Borges are significant examples of short story authors. Also, literary journals with international circulation, such as

Ford Madox Ford's *Transatlantic Review*, *Scribner's Magazine*, and Harriet Weaver's *Egoist*, provided a steady and prime exposure for young writers.

The short stories of Ernest Hemingway, for example, may often gain their force from an exploitation of traditional mythic symbols (water, fish, groin wounds) or are structured around a psychological, rather than physical, conflict. As well "The Dead," the final story in James Joyce's *Dubliners* (1914), builds from a casual mention of death and snow early in the story to a culminating paragraph that links them in a profound vision.

As the familiarity with it increased, the short story form itself became more varied and complex. The fundamental means of structuring a story underwent a significant change.

Science Fiction

Fiction is a narrative that strings together events that are imaginary, not factual. A work of fiction can be a book, a play, an opera or a film. Although there is little consensus as to what the exact elements of fiction are, there are some basic ways to identify concepts within fiction and analyze their meaning. Fiction is highly subjective, so your analysis could be different from someone else's interpretation.

The earliest short fictions were accommodated within the ready-made narrative frameworks of the anecdotal traveler's tale, the dream story and the moral fable, sometimes embedding painstaking attempts to dramatize philosophical propositions within frameworks that had usually been employed for more frivolous endeavors.

Specifically, Science Fiction is a genre that explores the influence of science and technology on people and society. It does not describe science but explores imaginary, future scientific and technological developments. Science-fiction stories are often set in the future, and predict how technology will change the world of modern society (hard science fiction). This type has often successfully predicted real scientific developments or technical discoveries. For example, the possibility to walk on the Moon, predicted by H. G. Wells in 1901, or an online network like the Internet, predicted by A. Clarke in 1962.

Some other science-fiction stories are set in an alternative parallel time and explore a "what if" situations, starting from how society works and in commenting on human behavior (soft science fiction): for example, if Martians landed on Earth, or if people will build a reading mind machine.

Early science fictions can be the argument in favor of the Copernican theory of the solar system advanced by Johannes Kepler's dream story *Somnium* (1634), which includes an ingenious attempt to imagine how life on the Moon might have adapted to the long cycle of day and night. As well, Voltaire's *Le Micromégas* (1752) employs a gargantuan native of Saturn to pour witty but devastating scorn on human delusions of grandeur.

Ideas about the possibility of flying machines, alien races, and advanced civilizations have existed throughout the history of literature. For example, there are also ancient Arabic or Hindu stories talking about space travel or submarines, as well as Shakespeare or Dante themselves who wrote about fantastic worlds.

Cyrano de Bergerac wrote about traveling to the Moon in 1656, as well as Jonathan Swift about alien cultures in *Gulliver's Travels* in 1726. Mary Shelley's *Frankenstein* (1818), classified as a horror story, really was an extraordinary example of a science fiction, based on scientific principles, exploring the consequences of pushing science beyond its normal limits.

Jules Verne (*Journey to the Centre of the Earth*, 1864; *From Earth to the Moo*n, 1865; *Twenty Thousand Leagues Under the Sea*, 1870) and H. G. Wells (*The Invisible Man*, 1887; *The Time Machine*, 1885; *The War of the Worlds*, 1898, on a Martian invasion of Earth; *The Island of Dr. Moreau*, 1896, on animal experiments and vivisection) were probably the two most influential early science-fiction writers. *Astounding Science Fiction* (founded in 1930) was one of the magazines that influence the diffusion of Science Fiction, starting a Golden Age of the genre.

New political and scientific topics were described, such as secret groups, police states, alternative realities, the question of mind control, artificial intelligence life, nanotechnology, and the influence of the Internet on the human life or supernatural powers.

Science Fiction in Didactics

The danger of making Science Fiction (SF) a didactic genre is a debated question.

First, there is the problem that the cited science could become obsolete. A scientific theory that underpins a story in one decade could be completely disproven by the next. For example, when Mary Shelley's *Frankenstein* was written, the precise nature of electricity was still under debate, and many scientists believed that it was a life-giving fluid that really could reanimate dead tissue. It stays a powerful novel about hubris and the human consequences of scientific experimentation. Similarly, Edgar Allan Poe wrote his most confronting SF when contemplating the subject of mesmerism, which still existed in the grey area between science and spirituality in the 1840s.

Science itself continuously is in a state of flux, as it separated from older traditions of natural philosophy and allied itself more closely with the materialistic discipline that society recognizes today. Just as scientists were testing those pseudoscientific ideas, so too were authors experimenting with them in fiction. It was all part of a wider tapestry of accepting those strange ideas into both science and society.

The level of expected plausibility can also vary depending on what subgenre the reader is dealing with. Since the 1950s, in the USA the term "sci-fi" was used to define low-quality science-fiction stories and low-budget science-fiction films. Nowadays, serious science-fiction authors refer to the genre with the initials SF.

The reader expects the stories of Hard SF, for example, to be generally accurate in their representation of science. However, SF stories can be indicative of the age in which the reader lives. For example, the anti-vaccination lobby is prominent in our media, and SF has a responsibility to explore that debate, to capture a snapshot of our fractured society.

The genre has always been a mirror of our world, and the mirror must sometimes reflect the unpalatable ideas alongside the good. Works of fiction may not have a responsibility towards scientific accuracy, but they do have a responsibility to challenge our preconceptions of the world.

On the other hand, while a general narrative thinking is focused on intentions and actions of people and is strongly anchored to a contextualized experience, a paradigmatic or logical-scientific thought expressed with formal and scientific descriptive modalities. It does not accept a "belief", looks for the foundations, the possible justifications and the implications for measuring the plausibility of ideas. It is based on categorization and conceptualization, according to a bottom-up path, namely from the particular to the general, tending to the progressive abstraction, and the procedures of elaboration and empirical demonstration are identified with the principle of coherence and non-contradiction.

This work aims to be an advanced form of dissemination of scientific knowledge and research in an accessible and easily understandable form. Scientific writing is often belied as dry, stale; a block of indomitable, indecipherable text buttressed with vague language, passive voice, countless equations, and overly constructed, borderline-run-on sentences that seem to drag on and on and on.

Peer-reviewed scientific discourse is often viewed as a special form of communication, exempt from the qualities of narratives to which humans inherently relate. However, our findings support an alternative interpretation: scientists can engage readers and increase uptake by incorporating narrative attributes into their writing styles (Hillier et al., 2016).

This is a work for Academic use, but it can become an excellent textbooks for Upper Secondary School (Level-A in the UK).

ORGANIZATION OF THE BOOK

This book consists of 10 chapters. Each chapter contains:

- A scientific introduction
- A didactic introduction to the science-fiction story
- The short SF story
- Suggested standard didactic activities
- An extended analysis of the main scientific topic
- A conclusion

Each series contains believable characters in realistic situations; characters that young people can identify with. In each series, there are many human characters, at least one robotic vehicle, and at least one major AI that appears only in virtual environments.

The topics presented are current and relevant to both a college and high school setting. The manuscript clearly demonstrates the power and utility of academic intersectionality between science and literature. This is an effective approach to understand the issues of science, historical events and literary devices (e.g., narrative, storytelling, etc.).

Through the organization of the book and the layout (e.g., premise, background information, how the chapter works and didactic exercises, etc.), all contribute to the strong presentation of key issues, problems, trends, and particularly new ideas and innovation emerging science to the reader.

Furthermore, there are questions and explanations of terms and concepts at the end of each chapter in order to probe further thinking.

Detailed Table of Contents

This book is divided into three sections each addressing a major problem of the 21st century. The individual stories then depict the experiences that people could have in working some aspect of a solution to that problem. Some of the people are artificial intelligences (AIs); well, welcome to the 21st century. Teaching materials are included with each story to make up the chapter.

A brief description of each of the chapters follows:

Section 1: The Big Moon Dig Stories

The first section, "The Big Moon Dig," is a series of three stories about young people participating in a grassroots effort to continue human space exploration. Depicted is a young people's movement exploiting 21st century technology to continue human movement out from Earth and in the process again find the forward-looking, positive vision of the future that was once provided by Apollo to the Moon.

Science: The key scientific ideas for the Big Moon Dig series are:

1. **Lunar Environment:** The surface of the Moon is a harsh environment with health-ruining radiation; hard vacuum, severe temperature swings, and pelting by small meteorites. Furthermore, nearly the entire lunar surface is covered with a finely ground rock called regolith that grinds at seals and moving parts. Designing a place for humans to live in such an environment is a major challenge.
2. **Construction of a Lunar Settlement:** A human lunar settlement must provide protection from (1) radiation, (2) temperature extremes, and (3) micrometeorites. The design depicted in these stories involves burying inflatable habitats under several meters of regolith. This effort then provides the name for these stories, the Big Moon Dig.
3. **Organization of a Grassroots Human Space Program:** Funding for space settlement primarily from governmental sources does not look plausible. Modern communication methods might now allow the organization of a human space program organized from the grassroots up. This process is here called Massive Online Vetted Expedition (MOVE).
4. **Humans and Artificial Intelligence:** Exponential advances in AIs are expected with the induction of specialty chips sets very soon. These stories show how one possible arrangement of the human/AI interaction might then develop.

Chapter 1: Consideration of a Grass-Roots Space Program, a Didactic Introduction – Story: "The Big Moon Dig"

Apollo to the Moon was a great adventure and a great story. However, there are many problems on Earth now that are only newly understood (global warming, etc.) and resources for human space exploration are consequently scarce. It is now highly unlikely that the resources for a grand space challenge, like going on to Mars, will be available for some decades.

The 21st century is not the 20th. Great computational and communication power is now directly in the hands of people. New types of collective actions are now possible. Could these new powers be organized for a 21st century style human space expedition?

Plot: The first story in this series is also called "The Big Moon Dig", and is about a team of Earth-bound humans, the Rocky Horror, supporting an AI driven lunar excavator, Digger03, in the early construction of a lunar settlement. This day the work is halted for the controlled explosion of a large bolder. Everyone on the project attends in the virtual reality version of the project. One of the long established members of the team takes this opportunity to announce that she is leaving the project and everyone is upset by this news.

What would it be like to work in a grassroots space program that that let, no that would only succeed if, masses of people were to participate? What would it be like if a few of those people were artificial intelligences?

The organization of the effort, called Massive Online Vetted Expedition (MOVE), is described through the interactions of the members of this team. One member of the team, Digger03, is a lunar excavation machine controlled by an AI. In addition to the limited electronics it has built into its lunar robotics, it has additional electronics on Earth that give it limited capability to function as a person.

The action of this story jumps from the physical location of the settlement construction near the Moon's South Pole (Real Moon); to a large virtual space locked to the physical site where the expedition participant's interface avatars, including those of the AIs, can interact (Virtual Moon); and to locations on Earth (Earth) where the participants interact electronically.

The action of the story is about one person leaving the team and a new person taking her place. All the participants have spent thousands of hours in training, simulations, and then in actual participation in the expedition. They are upset by the sudden change that is occurring and react in different ways: some happy, some sad, and the AI is indifferent. This action occurs in the middle of a big explosion party.

The questions presented by the Big Moon Dig story include:

1. What would a grassroots space program be like?
2. Would there be people who would be willing to put in the time and commitment required?
3. Would such an effort find again the positive vision of the future that Apollo to the Moon generated?

Society needs positive vision, particularly for young people, so that they can take on big problems. With such a vision, solutions to most of Earth's current problems become possible.

Chapter 2: The Humanity of an Artificial Intelligence – Story: "Stanford's Digs"

In the Big Moon Dig stories, people want so much to participate in human space exploration that they will devote years of their lives to training and then days to the boring tasks of actually operating a lunar rover. These people clearly have purpose.

Plot: In the second story in this series, "Stanford's Digs", a big Earth-based AI (Stanford01) adopts the purpose of space exploration all by itself and then it wants in on the current space action. It soon discovers a way in by befriending a lessor AI, the lunar excavator, Digger03. Some people are excited; some people are upset; while the big AI does what it needs to do to get its foot in the space door.

If a person tells an AI to do something, the AI then has a task and with it a purpose too. The question in this story is can an AI take on a new purpose all by itself.

If that new purpose of the AI were space exploration, would people object? Remember they spend years in training for their positions. What manner of tricks would the AI have to use to fulfill its new purpose? Would that be fair? Would it be ethical?

In all these stories, the major AIs are fully incorporated. This gives the AI a whole list of rights and responsibilities as a corporate person. Incorporation also gives it a board of directors that can provide some human control over the AI. The key element of this story is a fight over control of Digger03's board of directors.

The reader may ask:

1. Is there a meaningful difference between human space exploration (using all manner of tools including AIs) and Human/Machine space exploration where all work in a symbiosis?

2. If AIs and people work together, who should lead the way?
3. What level of control should humans have over the AIs? Is it simply the old Master/Slave, or is there possibly a Human/Machine Symbiosis?

Society needs many technology advancements to address the great problems of the 21st century. AIs that are both powerful and cheap are probably a key one of these innovations. This is exactly the advancement promised by the makers of the new specialty chip sets due out shortly—AI on a chip. However, how can society employ AIs powerfully while maintaining a productive level of control?

Chapter 3: AI and the Movies: A Storytelling About Communication – Story: "Caterpillars on the Moon"

Plot: In the third story in this series, "Caterpillars on the Moon", a section of inflatable habitat has been landed near the settlement site and must now be moved into position. The section is designed to craw like a caterpillar by inflating bags on its underside. The lunar excavation vehicle, Digger03, is steering the section but runs into technical difficulties resulting in delays. Once the problem is addressed, a major celebration is called in which the new space-based major AI, SallyRide01, is introduced. Two Earth-side major AIs who both helped design SallyRide01 then squabble over her parentage.

This story looks at the technical difficulties that must be overcome to build a space settlement where humans can live long term. At the stage in the construction covered in this story, a new habitat section has been safely landed, has been unpacked, and now must be moved into place. Landing the section directly into the trench would simply be too dangerous. The few excavation robots on site are not powerful enough to carry it. Sending heavy equipment from Earth would be excessively expensive. The solution depicted here is to let the section crawl like a caterpillar on inflatable bags.

The reader might consider what other animal models might be used: a centipede with a dozen small legs, an elephant with four post legs?

The delay between a command sent from Earth and that command executed on the Moon is several tens of seconds at the best. A lot could go wrong in this amount of time. Early in the building of a lunar settlement, a substantial computer and AI will be needed in situ, on the site.

Designing anything this complex for operations in space is no small task. The efforts of the big Earth-side AIs will certainly be needed for this design work. Could this lead somewhere unexpected? Will these AI then think of the new lunar AI as their offspring? Is there a story to be found in such a squabble?

Society must find new ways to work hard problems in the 21st century. AIs are clearly a key part of this effort.

Section 2: Iron Seas Stories

Back on Earth, the 21st century is bringing many new problems (global warming, sea level rise, etc.). Can the processes and vision from the Big Moon Dig stories be applied on Earth?

Science: The key scientific ideas for the Iron Seas series are:

1. **Seeding the Oceans With Minerals to Sequester Carbon:** Under just the right conditions, microscopic marine animals will build skeletons of calcium carbonate that settle to the bottom taking enormous amounts of carbon out of circulation. Seeding the oceans with minerals could greatly increase this effect but maintaining the balance is very tricky (Powell, 2007).
2. **Humans and Artificial Intelligence:** Exponential advances in AIs are expected with the induction of specialty chips sets very soon. These stories show how one possible arrangement of the human/AI interaction might develop. Here major capital assets, such as a ship, are permanently associated with their own AIs.

About half the excess carbon dioxide generated by people ends up in the oceans. Under just the right conditions, microorganisms in the oceans can tie up that carbon dioxide in their skeletons (Powell, 2007). Again, under just the right conditions, those skeletons can settle to the bottom and build up to form limestone; think of the White Cliffs of Dover. This process takes the excess carbon dioxide out of the environment for millions of years.

The second series of stories, Iron Seas, looks at the possibility of fertilizing the oceans with just the right minerals, mostly iron, to promote the growth of just the right microorganisms in just the right places. Promoting the wrong organisms could result in catastrophes like red tides with large fish kills. If the work is done in the wrong place, where the water is too deep, the pressure will simply force the limestone back into suspension and there will be no buildup. Done right, it takes a surprisingly small amount of material added to make a big difference.

The answer proposed in these stories envisions a fleet of converted pleasure sailing yachts working as scientific vessels. This puts teams of people right on the water just to be sure that everything is going right. Of course, in these stories one of the key team members is the AI that is the boat.

Chapter 4: Technical Measurement, Control, and Forecasting of the Climate Change – Story: "The Fid"

Most people are not yet taking the problems of the 21st century to heart Most people have not bought into the need for their personal action. Most people are not yet in action.

Plot: This the first story in this series, "The Fid", looks at a time early in the project when a major piece of art, The Fid, and a supporting video environment are built to help people understand the 21st century problem of sea level rise. Considerable effort is needed just to protect the project from attacks by climate deniers.

Could art be used to communicate the passion needed as well as the technical knowledge? Could art generate buy in for dedicated, heroic action on these great problems?

Alternatively, should the whole effort be a game? One gigantic enthralling game that provides the experience and technical training that people need to envision themselves succeeding even if they are only there to have fun.

Does a person who provides the vision and artistic skill early in the work deserve remembrance in the prolonged effort? Is there a story in early development of a seed of a heroic effort? Alternatively, is the story the game? Or, is the story the people in both game and art?

Chapter 5: The Greenland Ice: A Story of Changing Climate – Story: "The Cabin Boy's Tale"

Human society has done great damage to the oceans of Earth. Are there things that can be done to address these problems? It will take a great many people to take on the great problems of the 21st century; could some of these people be valent young people? Could some of these people be readers of these stories?

Plot: A young person, Jace, spends a summer on a research vessel, the Yvette A. Wight, that is participating in a project to fertilize the oceans to sequester carbon from the atmosphere. Through interactions with the crew, and the sea, he learns more than he bargained for.

Surprisingly small teams of real people are most effective at doing many of the efforts needed to address major problems, but they must be right in the thick of it. Always expecting other people to do the hard, often dirty, work is just not good enough. With the right training, young people can be productive members of the teams.

After all, the world that the young people will inherit is in play. Is there a story in these young people's effort? Could that story be the reader's story?

Chapter 6: Ocean Acidification: A Problem for Policymakers – Story: "The Captain's Tale"

There are technological solutions to the problems of the oceans. Action is possible. Nevertheless, whatever is done it must be watched very carefully. This monitoring effort will take big data taken from space and from the air, but it will also take a major hands-on effort by people right at the water's surface.

Plot: The normal operation of the Yvette A. Wight, is interrupted by a major breakdown in the process in the form of a red tide. As the whole project is controversial, the crew must take the scientific data needed to understand what went wrong and then get that data to the big lab on shore.

Every effort to address Earth's environmental problems will be controversial. Some people will want to do everything possible. Others will want to do little and let the Earth heal by itself. No set of actions could possibly please everybody.

Sometimes things will go wrong. If a great many people are watching exactly what is being done when something goes wrong, then a great many people will want to argue about what to do next.

Is there a story in this effort to get things right? Is there a story in the conflict that occurs when things go wrong?

Section 3: The Herd Stories

The great problems of the 21st century cover both land and sea. There are new management practices for the land that must be tested on a large scale. Many possible approaches must be worked out by the hard work of doing things by real people.

Science: The key scientific ideas for the Herd series are:

1. **Reestablishing the Symbiosis Between Herd Animals and Grass:** Human actions have let enormous areas of grassland degenerate into

deserts. Now large-scale tests (Savory, 2013) have shown that the grasslands can be reestablished by reestablishing the symbiosis of grass and grazing animals. This action will sequester a large amount of carbon in deep rich soils.

2. **Humans and Artificial Intelligence:** The AI in these stories are based on the concept of Human/Machine Symbiosis and not on the more common Master/Slave. This design places a premium on humans and AIs forming teams. These stories show one possible arrangement of the human/AI interaction.

Most ecosystems of the Earth are made up of symbiosis, multiple species supporting each other. For example, most of society's food crops are in symbiosis with the fungus that make up rich soils.

When people damage these symbioses, entire environments can degrade beyond use. This is exactly what has happened in many cases of desertification. The poor use of herd animals has damaged the grasslands where the herds once prospered.

The ruminate animals evolved in symbiosis with grass. The grass evolved its growth buds down in a structure right near the ground. Having their leaves eaten off simply does not hurt the plant all that much. The herd animals evolve big, complex stomachs that could digest cellulose. The hard feet of the animals selectively trampled the other plants, like tree shoots, leaving the grass alive while breaking up hard surfaces. Over time, grasslands, the savannas, became one of the most common ecosystems of Earth. Consider the Serengeti with its great herds, and the central plains of North America with its bison.

Reestablishing the symbiosis of herd animals, in this case cattle, and grasslands has been successfully demonstrated all the way up to some very big ranches (Savory and Butterfield, 2016). The secret is to restrain the animals to a small patch of grass for a limited amount of time and then move them on. Keeping the herd at just the right density insures that they eat all the available food without wasting very much. Moving them on at just the right speed insures that their urine and waste will water and fertilize the grass. The herds then move in ever-restless great cattle circles.

Once the symbiosis is reestablished, the grass will return even in very dry country. Monitoring this effort on a large scale will not be easy. Big data from satellites will help, but ground truth readings right in the middle of the herd are critical too. This series of stories features an AI controlled all-terrain vehicle, Dusty04, that does most of the monitoring work and a young woman, Zane, whose life is tied up with these cattle circles.

Chapter 7: In Action on Desertification: A Didactic Introduction – Story: "The Matriarch"

The key to the success of the great cattle circles is water. The inside of a ruminant's stomach is a very wet place. In dry country, deep wells are needed at critical points along the cattle circles. The cattle will not need nearly as much water as a large irrigation project would require so there will be enough water at the new wells for a village with its kitchen gardens.

Plot: An autonomous all-terrain vehicle, Dusty04, is checking out a wellhead before the thirsty herd arrives. A matriarch elephant moves out of the bush having smelled the water. Even this small group of elephants could destroy the wellhead if upset. The AI decides to try to communicate with the lead elephant offering to let them drink if they will then move off. A woman from the village, Imani, reluctantly supports this effort but is very much concerned as elephants and the village's crops simply do not mix.

The people's crops will have to be synchronized with the passing of the herds so that the animals can feed on the crop stubble and fertilize the ground. In fact, the life of the village will be fully tied to the cycles of the herd movements. These cycles will strongly effect the young people of the village as they grow and develop.

Two types of AIs appear in these stories. A medium AI closely monitors the herd. The key example here is Dusty04, who is an all-terrain vehicle. The larger AI, like the Janets, are physically in data centers far away but can appear as interface avatars on virtually any screen. They handle the big data including the satellite images and provide management. Both types of AIs are independently incorporated so they are treated as corporate persons in these stories.

The question of who is a person is a major theme in this story. The medium AI, Dusty04, has an encounter with a matriarch elephant. As Dusty04's purpose includes monitoring the local wildlife, it tries to communicate with the elephant using what sounds and gestures are then known. If Dusty04 is a person of a sort, is the elephant a sort of person too?

The efforts to address the great problems of the 21st century will affect many people's lives. Some will have to move to new locations; some will have to end practices that their ancestors have used for many generations. In many ways, this will be a hardship, but in other ways it will be an opportunity, particularly for the young people.

If an AI can be a person, how about an elephant, or a whale, or a bonobo? Who gets to make this decision?

Is there a story to be found in humans and AIs forming a symbiosis together to take on great problems? Is there a story in the possibility of AIs interacting with the other animals on this planet too?

Chapter 8: Climate Protection and Ecosystem: A Storytelling to Train Awareness – Story: "The Triceratops"

The herd animal / grass symbiosis evolved long before there were cattle. There is good evidence for its existence in the last period of the dinosaurs, the Cretaceous. The herd animals then included early members of the Triceratops lineage.

Plot: in this story, "Triceratops", the AI, Dusty04, conducts a training session for the village near the pumping station. A major video game is used to find likely trainees for cattle work among the young people of the village. Dusty04 adopts a young girl, Zane, due to her interest in AIs and marks her for future special training.

It will take a number of trained people to keep the crop circles functioning properly. In rural areas such jobs can be very important. In these stories, a great video game is used to train and select local candidates for these jobs. These training exercises are run by the AIs.

The AIs run a very popular game that provides both training and is a demonstration of skills. There is much competition over the head sets needed to play this game among the village young people.

In the story, one young person, Zane, stands out to the AI. She is much too small to competitively hold on to a headset for the game, but an AI notes her positive attitude and interest just the same.

If AIs and people are working together in symbiosis, does the AI having this much power over the young people's future seem fair? Should the one with the most technical knowhow make that decision, whither human or AI? Should there be a governing body to make such decisions?

If the AI sees a need for trained people in the future, how much power should it have over the training of people to fill that slot? Would the reader accept career advice from an AI? Is there a story in the control of people's lives by the AIs?

Chapter 9: AI and People: A Storytelling About Interactions – Story: "Seascape With Sharks"

If AIs are in a symbiotic relationship with humans, and everyone is working together to solve big problems, how much power should the AI have at stopping one human being from taking advantage of another? Should an AI have important law enforcement powers?

Plot: A local port is clearly being flooded out by sea level rise. A major AI, JanetM, has conducted a technical study to site a replacement harbor under the political direction of a judge. A seedy human, the Shark, sees this as an opportunity to make big money off the project if he can disable the AI long enough to get himself situated as the representative of the local people who are entitled to compensation. The Janets do not go down without a fight.

It is fine to talk about everyone working together but sometimes hard decisions have to be made. Often such decisions are good for the whole but are clearly to the disadvantage of a few. Often people know there is a problem but are simply not ready to jump to a difficult solution.

If an AI plays a critical role in making the decision on an approach to a problem that will change people's lives and is the one with the technical data, should it be the AI that presents the solution to the people affected? Should the AI have power to require the technical solution that AIs have worked out?

If one human being is clearly taking advantage of others and attacking the AI, should an AI be allowed to run a vendetta against that human being? Who gets to decide if the human has done wrong? Is the story in the actions of the AI working the problem or in the emotional response of the people effected, or is it in the law enforcement actions of the AI?

Chapter 10: AIs Training People: A Storytelling about Building Life Teams – Story: "The Girl at the Pump"

AIs do not have children in the same way people do. If a powerful AI takes a special interest in a human child and guides the child's development and education, does that make the AI a stepparent? Is that even ethical? Is it fair to the other children?

Plot: A young person, Zane, has been trained for years by a major AI, JanetA, to be a project logistics manager. Despite having never been more than 40 kilometers from her remote village on a cattle circle, she jumps at the chance to support an expedition even if it is headed to the artic.

If an AI sees a need for a person with specific training and follows a child for years seeing that that young person gets training for a position that the AI needs, is that fair to the child? If a young person decides that they simply must leave the area where they grew up, should an AI help them to go? Should the reader seek the aid of an AI to do what is in its heart even if it is not what its family expects of their child? What if this leads the young person to traveling far, far away from its family?

Is the big story here the interaction of the AI and child, or is it the coming of age of the young person?

REFERENCES

Altet, M. (2009). *L'université peut-elle vraiment former les enseignants*. De Boek.

Aylett, R. (2006). And they both lived happily ever after? In G. Dettori, T. Giannetti, A. Paiva, & A. Vaz (Eds.), *Technology-mediated narrative environments for learning* (pp. 5–25). Rotterdam: Sense Publishers.

Bruner, J. (2003). *La fábrica de historias. Derechos, literatura, vida*. FCE.

Bruner. (1990). *Acts of Meaning*. Torino: Bollati Boringhieri.

Damiano, E. (2007). *Il sapere dell'insegnare. Introduzione alla Didattica per concetti con esercitazioni*. FrancoAngeli.

Damiano, E. (2010). *La mediazione didattica*. Per una teoria dell'insegnamento.

Damiano, E. (2013). *Per una teoria dell'insegnamento*. Franco Angeli.

Hawken, P. (2017). *Drawdown: The Most Comprehensive Plan Ever Proposed to Reverse Global Warming*. Penguin Books.

Herman, K. L. (2003). Predictors of Success in Individuals with Learning Disabilities: A Qualitative Analysis of a 20-Year Longitudinal Study. *Learning Disabilities Research & Practice*, *18*(4).

Hillier, A., Kelly, R. P., & Klinger, T. (2016). Narrative Style Influences Citation Frequency. *PLoS One*, *11*(12), e0167983. doi:10.1371/journal.pone.0167983 PMID:27978538

IPBES Intergovernmental Science-Policy Platform on Biodiversity and Ecosystem Services. (2019). *Nature's Dangerous Decline 'Unprecedented', Species Extinction Rates 'Accelerating'*. Retrieved from https://www.ipbes.net/news/Media-Release-Global-Assessment

Jackson, T. (1995). *European management learning: A cross-cultural interpretation of Kolb's learning cycle*. MCB UP Ltd. doi:10.1108/02621719510086174

James, C. H., & Minnis, W. (2004). Organizational storytelling: It makes sense. *Business Horizons*, *47*(4), 23–32. doi:10.1016/S0007-6813(04)00045-X

Laybourn-Langton, L., Rankin, L., & Baxter, D. (2019). *This is a Crisis, Facing up to the age of environmental breakdown.* Retrieved from https://www.ippr.org/files/2019-02/this-is-a-crisis-february2019.pdf

Lesh, R., & Doerr, H. (2003). Beyond Constructivism. London: Academic Press. doi:10.4324/9781410607713

Lieblich. (1998). *Narrative Research: Reading, Analysis, and Interpretation*. Sage Ed.

Luckin. (2001). Narrative Evolution: Learning from Students' Talk about Species Variation. *International Journal of Artificial Intelligence in Education*, 12.

McEwan, H., & Egan, K. (Eds.). (1995). *Narrative in teaching, learning, and research*. New York: Teachers College Press.

Mezirow, J. (1991). *Transformative dimensions of adult learning*. San Francisco, CA: Jossey-Bass.

Nonaka, I., & Takeuchi, H. (1995). A theory of organizational knowledge creation. *International Journal of Technology Management*, *11*, 833–846.

Polkinghorne, D. E. (1989, July). Narrative Knowing and the Human Sciences. *Review by: Robert J. Richards American Journal of Sociology*, *95*(1), 258–260.

Powell, H. (2007). *Fertilizing the Oceans with Iron, Should we add iron to the sea to help reduce greenhouse gases in the air?* Retrieved from http://www.whoi.edu/oceanus/feature/fertilizing-the-ocean-with-iron

Ricoeur, P. (1981). *La métaphore vive, Seuil, Paris 1975, trad. it. La metafora viva. Dalla retorica alla poetica per un linguaggio di rivelazione*. Milano: Jaca Book.

Rossi, P. G. (2010). Metodi di indagine per analizzare le modellizzazioni degli studenti e le modellizzazioni degli insegnanti. In B. D'amore & S. Sbaragli (Eds.), Matematica ed esperienze didattiche. Bologna: Pitagora.

Savory, A., & Butterfield, J. (2016). *Holistic Management, A Common Sense Revolution to Restore our Environment*. Island Press.

Scalise Sugiyama, M. (2001, July). Food, Foragers, and Folklore: The Role of Narrative in Human Subsistence. *Evolution and Human Behavior*, *22*(4), 221–240. doi:10.1016/S1090-5138(01)00063-0

Schank, R. (1990). *Tell Me a Story: A New Look at Real and Artificial Memory*. New York: Charles Scribner's Sons.

Schön, D. A. (1987). *Educating the reflective practitioner*. San Francisco, CA: Jossey-Bass.

Wertsch, J. V. (1998). *Mind as Action*. Oxford University Press.

Acknowledgment

The authors would like to thank the early members of The Big Moon Dig:

Kenisha Eberhard: She provided many of the illustrations for the Herd series and much of the insight into the needs of young people needed to write these stories.

Kent McCullough: He provided many of the technical details and his emotional support that kept the Big Moon Dig going for years.

Section 1

The Big Moon Dig Stories

The first section, "The Big Moon Dig Stories" is a series of three stories about young people participating in a grassroots effort to continue human space exploration. Depicted is a young people's movement exploiting 21st century technology to continue human movement out from Earth and in the process again find the forward-looking, positive vision of the future that was once provided by Apollo to the Moon.

Science: The key scientific ideas for the Big Moon Dig series are:

1. **Lunar Environment:** The surface of the Moon is a harsh environment with health-ruining radiation; hard vacuum, severe temperature swings, and pelting by small meteorites. Furthermore, nearly the entire lunar surface is covered with a finely ground rock called regolith that grinds at seals and moving parts. Designing a place for humans to live in such an environment is a major challenge.
2. **Construction of a Lunar Settlement:** A human lunar settlement must provide protection from (1) radiation, (2) temperature extremes, and (3) micrometeorites. The design depicted in these stories involves burying inflatable habitats under several meters of regolith. This effort then provides the name for these stories, the Big Moon Dig.
3. **Organization of a Grassroots Human Space Program:** Funding for space settlement primarily from governmental sources does not look plausible. Modern communication methods might now allow the organization of a human space program organized from the grassroots up. This process is here called Massive Online Vetted Expedition (MOVE).
4. **Humans and Artificial Intelligence:** Exponential advances in AIs are expected with the induction of specialty chips sets very soon. These stories show how one possible arrangement of the human/AI interaction might then develop.

Chapter 1
Consideration of a Grass-Roots Space Program:
A Didactic Introduction

ABSTRACT

Human space exploration has historically provided a great many people with a positive vision of the future. At this time, society faces many 21st century problems (global warming, sea level rise, etc.) and could use some of that vision. The economic state of the nations that historically paid for this exploration does not currently allow for a large and expensive new space initiative, like Apollo to the Moon or a trip to Mars. Nevertheless, there have been great strides in computing and resulting social media. Could a very large number of dedicated people self-organize into a grassroots human space program? This story envisions such a movement and the lessons today's students could learn from the attempt.

INTRODUCTION

Twelve men walked on the Moon between 1969 and 1972. They left scientific experiments on the surface and brought back to Earth almost 400 kg of rocks and surface samples. What is more important is that they left us with a positive vision of our future and an historic since of accomplishment.

To face climate change now, we need to regenerate some of that positive vision of our future, but to do so on a shoestring budget as most of our

DOI: 10.4018/978-1-5225-8401-8.ch001

resources need to be committed to Earth's problems. Can young people develop a low-budget, democratic, grassroots plan to build a lunar settlement? Can the latest technological and social media developments be applied to this problem? Can humanity as a whole maintain a positive vision of our future in the face of whatever comes?

The answer to all these questions in this, "Big Moon Dig" series of stories is an enthusiastic Yes.

Topics for Discussion

The following discussion points come from information in this chapter:

1. Does the positive vision of the future provided by Apollo to the Moon have value today?
2. How has our knowledge of the Moon continued to build?
3. Why is it so hard for humans to live on the Moon?
4. Is competition for funds between exploration of space and addressing global warming helpful today?
5. Can humans and Artificial Intelligences (AIs) work together in teams today? In ten years?
6. Given modern social media, e-education, and AI, is a low-cost grassroots human-plus-AI space exploration now possible? In ten years?

Recent Progress on Lunar Exploration

In recent years, many countries have decided to resume the explorations of the Moon. Their sounders in orbit have explored the entire surface in great detail. An important discovery was the presence of water ice mixed into the ground rock at the bottom of the craters near the lunar poles. Protected from the sun's rays, this ice could have been there for billions of years. Water ice could be used by future lunar settlers to produce breathable oxygen, rocket fuel, and drinking water.

The first European lunar sounder was SMART-1, which orbited the Moon from November 2004 to September 2006. SMART-1 took detailed images of the planet's surface and studied the composition of the surface rocks. The mission ended with an intentional crash on the lunar surface.

NASA now has the stated intention to return to the Moon with the mission statement, "We will address the American space program to human exploration

Figure 1. SMART-1 sounder, by NASA (Source: https://www.esa.int/spaceinimages/Images/2007/02/SMART-1_mission_at_the_Moon)

and discovery, which means launching American astronauts beyond the Earth's low orbit for the first time since 1972. This means establishing a renewed American presence on the Moon, a vital strategic objective, and on the base for lunar exploration, America will be the first nation to bring humanity to Mars "(VP Mike Pence, Oct 5, 2017). Funding for this mission has however been minimal so the effort has been limited to early design and prototype testing.

Exploration Mission-1 (EM-1) will be the first integrated test of NASA's new deep space exploration systems: the Orion spacecraft, Space Launch System (SLS) rocket and the ground systems at Kennedy Space Center at Cape Canaveral, Florida. The Space Launch System (SLS) is a non-reusable heavy orbital launch system derived from the Space Shuttle and designed by NASA, the United States government space agency. Launches to test these systems are expected in the early 2020s.

Yet even well laid plans can hit hard times. For example, the Augustine Commission, established by the new President Barack Obama, deduced that several delays were necessary that postponed the flights to the 2020s (Kampler, 2019). Much of NASA's current resources are tied up getting the Web Space Tellescope completed and launched. This cost will last at least until the launch now scheduled in 2020. Major tax cuts by the Trump Adminstration farther reduce the federal funds that are possible for major space exploration misions and this Admistration has now put the emphasis on a new Military Space Force and away from human exploration.

Figure 2. Comparison between crews (Source: Volpicelli, 2015)

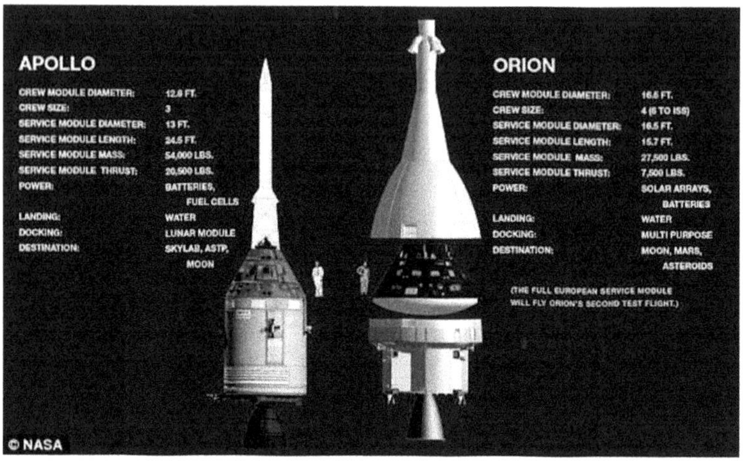

Figure 3. Bucket wheel moon excavator designed by Colorado School of Mines (Source: http://www.thelivingmoon.com/41pegasus/01archives/Bucket_Excavator_00.htm)

Efforts to return to Space do continue with the development of rockets of modern design by several commercial companies and many university projects to design critical components. Progress is slowed but has not stopped completely.

Historically Apollo to the Moon provided an enormous number of people with a vivid, positive vision of the future. Such a vision would be a major asset in addressing the great problems of the 21st century if it can be reestablished.

The possibility of a major nationally funded mission back to the Moon or on to Mars currently seems remote. However, modern advances in digital communication and artificial intelligence bring up the possibility of a grassroots movement to provide most of the manpower and support the market for the commercial ventures, while accepting the limits on governmental expense. The three stories in this series, The Big Moon Dig, explore this new approach.

Who Needs Vision?

The characters in all these stories have a purpose. They are trying to address the great problems of the 21st century (global warming, sea level rise, etc.). But why have the authors included this space opera and not other more direct efforts?

Back in the 1960s, critics of Apollo to the Moon and the space program in general complained about throwing money into space when there were so many problems right here on Earth that needed attention. Spend the money right here at home.

In the end, Apollo to the Moon answered that criticism through accomplishment. The people of Earth reveled in the excitement of placing a man on the Moon.

Space exploration has been a source of positive vision since the first science based science fiction stories of the late 1800s. That same positive vision is needed today to face the great problems of the 21st century. Reinvigorating the Apollo vision for all Earth's people is worth the price of at least a modest human space program even if a huge space program is priced out of the question.

How can Earth's people now recapture such vision? Would a grass roots, 21st century style space program do the job?

How This Chapter Works

This first analyzed story is entitled "The Big Moon Dig", and is the first story of the Big Moon Dig Series. This study focuses on four parts: an introductory didactical analysis, the Science Fiction (SF) story, a technical in-depth analysis, and a proposal of educational activities.

DIDACTIC ANALYSIS OF "THE BIG MOON DIG"

The story is analyzed as science fiction storytelling, highlighting how it can be used both to explore crucial science issues on the settlement of the Moon and improve computational thinking and creative problem-solving.

It has two basic components: the product (what) as a communicative act, and the process (how) to get results with different ways and effectiveness. Sight, hearing, touch, smell, and taste are based on memory, and they return lifelike in the narrative format, deriving from previous experience. It was opportunely observed that "There would be no memory without narration "(Fontana, 2009: 15).

What is the difference between the narration of a fantasy world in a wonderful past ("Once upon a time") and in a timeless world, and the narration of a possible real future, maybe with better life prospects or totally different from actual experience but with human relationships well known to the reader? Both are open to fantasy, but the futuristic ones trigger curiosity and mobilize change of something. In these stories, the author also introduces real technological elements and the truth of a world that has already developed the ability to shift between the virtual and the real.

Figure 4. What and how in the story in Big Moon Dig (original design by dall'Acqua')

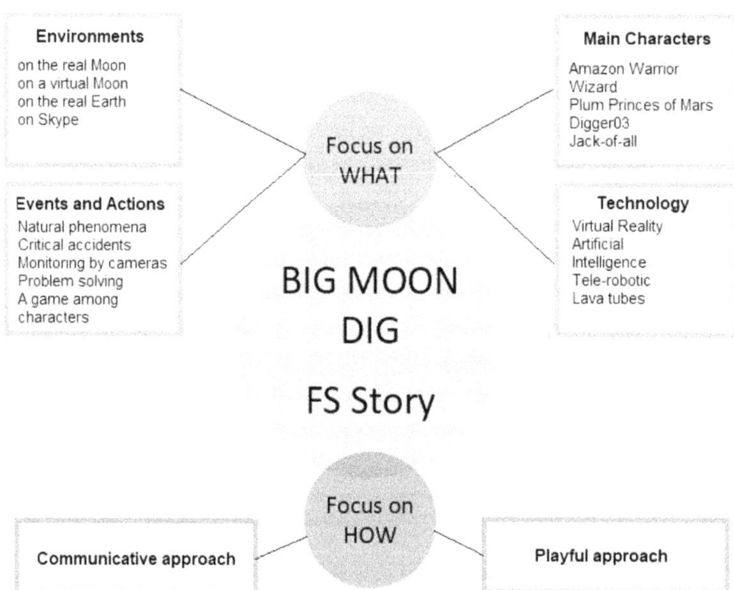

Environments

In this series, scenes take place sequentially but in three seperate environments: on the real Moon, on a virtual Moon, and on the real Earth:

1. **Real Moon:** The Real Moon is an identified superior construction site for a lunar settlement on the Moon, the Scott A Massif (key term). It is near the lunar South Pole and is near both water and power sources. Here many natural and human made phenomena happen. All these events are monitored by cameras. Here a few human-built machines do the real work of building a foundation for a lunar settlement.
2. **Virtual Moon:** The Virtual Moon is an enormous virtual reality environment and the primary setting for the action. This environment is locked to the physical reality of the Real Moon and provides a place for the interface avatars of the characters to interact. Here the full Rocky Horror team, including the interface avatars of the human members and an AI machine avatar, are intent on solving critical problems to overcome obstacles for the success of the mission. Many other avatars also may attend events on the virtual Moon. Physically this virtual environment is located in a major computing center in Nevada.
3. **Earth:** Earth is the location for the scattered human team members who rarely meet. They communicate electronically often over Skype.

Main Characters

On the virtual Moon the main characters are an Amazon Warrior; a Wizard, a Princess of Mars, a denizen of Sherwood Forest, and Digger03, a robot excavator. There they can interact and work out the many problems of building a lunar settlement without ever all being on the site.

1. **Amazon Warrior:** This interface avatar is run by a woman named Amelia. In this story, she leaves the project to find an excellent position in high tech industries. Her lunar experience is her primary qualification for her new job. Her sudden departure has upset most of the other characters.
2. **Wizard:** This character, also known as the old man, appears in several of the Big Moon Dig stories. He is a long-term member of the MOVE effort and the counselor of the Rocky Horror team. In real life, he is an IT professional. He often expresses the views of the author of the series.

3. **Plum Princes of Mars:** This is an interface avatar who is operated by a woman named Clarice. She is a foundation member of the Rocky Horror team. This presentation of a character is taken from the classic science fiction books of Edgar Rice Borrows that are set on Mars. This is also the basis of the costume worn by Princess Leia Organa of Alderaain in the Jubba the Hut sequence in the Star Wars franchise. As this is a famous science fiction character, many people would like to use it so each clone is given a different wardrobe color to distinguish it, hence the designation the color Plum here. This costume is completely impractical for any type of physical activity, especially fighting, and is simply for show.
4. **Digger03:** This robot/AI works as an excavator on the Moon. It is independently incorporated and so is a cooperate person (key term) with some, but not all, of the rights of a person. It does not have a gender and does not represent itself in virtual space as either sex and so is always referred to with the pronoun it. Its lunar AI is smart enough to keep the machine out of physical danger on the Moon, but not much more than that. It is electronics on the Moon are supported by a larger rack of electronics on Earth. It is a major character in the first three stories in the Big Moon Dig series.
5. **Jack-of-all:** This interface avatar is run by a major member of the Rocky Horror team and can take over for any missing member of the team on short notice, hence the name shortened from "Jack-of-all-trades". He appears as a denizen of Sherwood Forest complete with a long bow.

These characters form the Rocky Horror Team that operates Digger03 on the day shift. On Earth, they are the scattered volunteers in the MOVE project of the Big Moon Dig. They show the variety of people and machines that would be needed to build a settlement on the Moon.

Technology

The main cited technologies are: Virtual Reality, Artificial Intelligence applications, Tele-robotic operation, and space settlement. For a specific in-deph discussion, see the dedicated Technology Section below which includes supporting topics like lava tubes.

Communicative Approach

The communication process the oral language is of primary importance, as well as the starting point for learning a second language.

Stories motivate learning; they are fun and help develop a positive attitude especially towards the second language. The children are involved personally because they often identify with the characters and try to interpret the story and the illustrations thus favoring the development of creativity.

1. Listening to stories helps children pay attention to the rhythm, intonation, and pronunciation.
2. Storytelling facilitates the development of learning strategies such as listening to grasp the general sense, predicting later scenes, guessing meanings, and hypothesizing.
3. This story allows the teacher to introduce and revise words and structures, exposing the learners to the language and enriching their linguistic heritage, especially technical key terms.
4. In the case of a foreign language, the narration offers opportunities to present cultural information of the country in which the language is spoken. Every word has its own history, its particular meaning, its value that is not always easily translated into another language but that with the right material we can try to "pass" to the students, immersing them in the atmosphere of the second language.

Playful Approach

As R. Titone states "... play is the activity that most vitally and functionally promotes the total development of the child". (Titone, 2000) However, the emphasis on the game should not lose sight of the interests and abilities of students in the field of work that meets their needs to engage in something productive and important. It is therefore important to cut out moments dedicated to activities within the substantial class games and activities, capable of producing true emotional involvement, a fundamental engine of any form of learning

Approaching the Story

Apollo to the Moon was a great story. However, there are many problems on Earth now that are only newly understood and resources for human space

Figure 5. Digger03 on the Virtual Moon (original design by Riley)

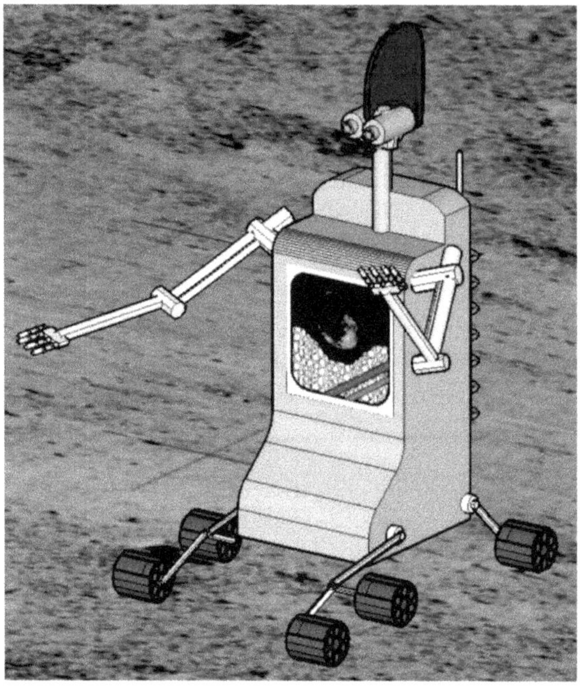

exploration are now scarce. The 21st century is not the 20th. Great computational and communication power is now directly in the hands of people. New types of collective actions are possible.

What would it be like to work in a grassroots space program that would only succeed if masses of people were to participate? What would it be like if a few of those people were artificial intelligences? What would their story be like? Could this turn out to be the reader's story?

THE FICTION STORY: "THE BIG MOON DIG" BY TOM RILEY

Premise

"Twelve people went to the Moon; then we stopped. Maybe, this time, it will take the efforts of a million people to get us back there."

On the Virtual Moon

This will be my last shift," said Amelia.

The Amazon Warrior spoke generally to the team but looked no one in the eye. They would be waiting right here for an hour, so it was as good a time as any to make the announcement.

"Isn't this kind of sudden?" the Wizard asked.

The team's counselor seemed to lean even heavier on his staff, with his misshapen hat dropping even farther down his forehead.

"True, sorry for the suddenness, but I have gotten a good job offer on the outside," the Amazon Warrior responded.

"I couldn't really talk about it. Then they suddenly insisted on an answer today."

"What's this you're saying?" the navigator, a Plum Princess of Mars, queried anxiously. "You're not really leaving, are you?"

The five members of the Digger03 Shift 1 Team, the Rocky Horror, stood in a group well away from the mass of other players gathering in the stands.

On the virtual Moon, they were an Amazon Warrior; a Wizard; a Princess of Mars; a denizen of Sherwood Forest, and a robot digger.

On the Real Earth

Physically the Virtual Moon is three racks of state-of-the-art electronic equipment in a major computer center in Nevada.

One of the racks contains dozens of blades with the latest deep learning chips and another one has all the high-resolution virtual reality hardware. The computer center provides power, cooling, and row-upon-row of routers to handle the heavy incoming traffic. The center also provides an advanced security firewall to protect the entire system.

The Virtual Moon's software generates an enormous virtual reality space that only vetted players are allowed to enter. Its foundation is the physical state of the settlement construction site on the Moon, updated minute-to-minute as the work progresses. Overlaying that are all manner of scientific and navigation aids that allow extensive modeling of all operations before they are physically executed on the real Moon.

The appearance of each player is, of course, that player's free choice. The Virtual Moon itself is personified by an interface avatar called the supervisor, but may be addressed directly at any time as "Computer".

On the real Earth, the four people in the Rocky Horror team have bland identities and are scattered across three continents. The Wizard is a senior IT professional; the Amazon Warrior is a recent liberal arts graduate; the Princess of Mars is a perpetual student; the denizen of Sherwood Forest is a martial arts instructor.

What they have in common is that they are all long established members of the Massive Online Vetted Expedition (MOVE), which is now breaking ground for the first human lunar settlement. Each team member has completed dozens of on-line courses culminating in certification as a remote robotics operator. Obtaining these certificates was their vetting; unvetted persons are only allowed to watch.

On the Real Moon

At the top of a rise at the very southern edge of the great Scott A Massif sits a box covered with blankets of metallic gold film. It is a communication hub, handling both local and Earth channels. Above the box is a single mast with a billboard sized photovoltaic panel that spins around once a month.

Topping the mast is an instrument box with an elaborate set of cameras. These cameras are the basis of the Virtual Moon; they are the sources of most of what the people of Earth now see of the settlement construction site of Armstrongville, Moon.

What the cameras see is the top of the Scott A Massif bathed in the harsh sunlight and long shadows that are characteristic of a near-polar location on the Moon. The Sun is well up now, a week into the lunar month, but only a few degrees above the horizon. It rarely ventures above 10 degrees at the construction site.

The Earth hangs in the black sky due north with its striking whites, greens, browns, and blues. It is always there and nearly in the same place, making a lazy figure eight over the course of a year. It never reaches more than two hands above the horizon but rarely touches it. Mother Earth is every MOVE team's best friend: always there, always comforting, always ready to communicate.

The Earth's disk appears much bigger than the lunar disk appears back home, almost four times larger. It goes through phases as the Sun travels on its circling path.

It is now a week after midnight, the Sun is due east, and the waning gibbous Earth gives a fine glow, throwing light at right angles to the Sun. That Earthshine lets the hub's cameras peer into many of the otherwise pitch-black shadows.

In places where the direct sunshine hits hour after hour, the surface temperature can reach 200 C. Conversely, in places where the sunshine is shaded from the surface and that are exposed to the cold blackness of space, the temperature can drop to -120 C. Fortunately, the regolith carries heat only very poorly so the temperature under as little as one meter of the covering of crushed rock is around 20 C and good for human habitation.

Despite the workable construction area here being both tens of kilometers long and wide, there are no truly flat spaces. Everywhere the surface slopes a little this way and that. The great flat areas of Earth, the planes and ocean abysses, were all created by liquid water. There has never been any liquid water on the Moon, ever.

Several billion years ago, this place was a great mountain range, part of a ring of an enormous basin, the South Pole – Aitken basin. Now every feature is worn into rounded shapes by so many eons of meteorite strikes.

Even the south edge of the massif is not a true cliff but rather simply slopes more and more until the wheels of a rover can no longer get traction in the powdered surface material and the vehicle will simply slide to oblivion into the cold and the dark.

The entire area is pockmarked with craters ranging in size from potholes to a house. None are especially deep but they turn the path for a wheeled vehicle into an obstacle course.

Across this landscape are scattered many rocks of all sizes, from pebbles to boulders. These erratics were blasted here by distant great meteor strikes and provide samples of the lunar crust from many locations and many ages.

The cameras also see four pieces of human-made equipment now on the Scott A Massif construction site. One immobile rover sits on a small rise with too many frozen wheel bearings, seized by prolonged exposure to the surface grit. Three working rovers still move slowly around doing the digging work.

Three others shoebox sized exploratory rovers have long since left the area to blaze trails down off the massif. Only their tracks leading off into the distance are still visible.

One of the functioning rovers is the solar powered excavator known as Digger03. It is a piece of custom lunar construction equipment, much less massive than one designed for Earth but still able to pick up loose dust and rocks and dump them in some other place, as long as it is operating in the Moon's 1/6 g and the rocks are not too big.

It is also an AI and a corporate-person; complete with some rights and an Earth-side board of directors. It is a semi-autonomous being supported by its three MOVE teams back on Earth; it is a direct descendent of the XPRIZE robots that brought live action back to the endlessly patient Moon.

On the Virtual Moon

"How long will it take to replace her?" Digger03 asked. Digger03 appeared as a half-humanoid robot and half dump truck with a tall neck, two camera eyes, and a solar-cell blue crest on its head. It had a boxy six-wheeled base and a backpack that was often full of rocks. When it became necessary, Digger03 was smart enough to take a safety action or ask a simple question on its own. How else could it learn?

"Only a shift or two at most," the Wizard replied.

"Jack-of-all can fill in for now. There is a long queue of qualified candidates for any lunar position that comes open."

On the Real Moon

The Scott A Massif is a plateau running east to west and standing about six kilometers high. It has local high points with sunlight for power about 95% of the time as its southern edge is a peak of eternal light, and it has continuous line-of-sight to Earth. It has a kilometers-wide buildable area and at least four routes that never exceed a 20% grade running a long way down to the permanently shadowed craters far below and to the south.

On a prominent rise sits the box of electronic equipment covered in gold blankets. Overlooking the dig site, it has a clear view of everything all the way to the horizon, even out into the dark and the cold that fills most of the 120 kilometers between the Scott A Massif and the lunar South Pole; but it can make out little out there in all the deep shadows.

This box contains two small telescopes that provide laser communication with Earth and several antennas for microwave, both to Earth and to the local rovers. It has a tall vertical mast with a billboard of a solar panel that it rotates slowly through the month to always face the Sun. Mounted at the top of this mast is a complex camera assembly with full track, pan, and zoom capability that are the foundation device for the Virtual Moon.

In the flat area to the west of the rise are two trenches. The first runs 30 meters north-south and lies complete and open. The new trench runs east-west and, at 20 meters, it is still being worked but is now approaching completion. It now ends abruptly with a large boulder; nothing will do but to blast that boulder.

Both trenches start with ramps, which showed a multitude of rover tracks running up to ground level and fanning out to various yards where the loose material from the dig has been dumped into piles. Useful looking rocks are sorted and laid out in a checkerboard pattern so they can be individually located should a particular rock later be needed for construction.

The entire surface of the site and, in fact, nearly the entire Moon is covered with fine-ground powdered rock call regolith. Millions of years of gardening by micro-meteorites have turned the Moon's top several meters into thoroughly broken rock. Its top six centimeters are very fine powder, then the deeper you dig, the bigger the erratic rocks get. The top layer of powder-fine grit is the dig team's worst enemy; it forever grinds away destroying mobile equipment's bearing and seals.

The plan for a magnificent future Moon settlement is now laid out in great detail. Key reference points are marked with piles of rocks called cairns. Human beings have marked their paths with stacks of rocks since the Stone Age; now the art of the cairn is the art of the Moon. Even a modest cairn on the Moon can easily last a million years.

Each cairn has a distinctive shape and can be considered a work of art. Some cairns have bits of color of Earth origin, perhaps a bit of gold blanket or a scrap of metal, but the bulk of the structures are stacks of natural rough-formed monotonous gray lunar stones piled as high as a Digger can reach. Deep in the foundation of each cairn is a flat rock with the names of the responsible dig team diamond inscribed on it.

There are no human footprints anywhere to be seen because no human being has set foot on the Moon in more than 70 years. What the cameras document is the magnificent desolation that is the Moon. To the south, they can see only the utter blackness of the cold and the dark where the promised water lies.

On the Virtual Moon

All the MOVE teams were gathering this day, hundreds of players and trainees. The amphitheater on the hillside below the communication box is now nearly full, holding a wide assortment of wizards, princesses, demigods, and trolls. It simultaneously presents a display of finery and a macho strut. Everyone is there both to see the blast and to be seen.

A handful of formally dressed outside dignitaries also fill a VIP box, looking rather out of place. One of these is an athletic lady with dark hair and brown skin. She is wearing a scarf of Spanish lace and a full skirt in bold rich colors that quite literally sings fiesta to you if you brush against it. Her nametag reads simply "JanetM".

On the Real Moon

Digger03 sits alone with its camera held high, resting in a slight depression. This gives it a clear view of the new trench running east to west before it, but provides some protection.

The big trench is 22 meters wide at the top with a V-shaped cross-section and a four-meter flat area at the bottom. The depth is a little over eight meters.

Decades before NASA missions had carried instruments specifically designed to measure the danger of space radiation to human beings. The results were not what the space enthusiasts had wanted to hear. The radiation danger for humans in space is much worse than had been foreseen or that anyone had ever included in their plans, about twice as bad or more.

The answer to this problem is to dig and dig deep. A shield of five meters of lunar regolith will provide about as much radiation protection as the Earth's atmosphere. That was considered acceptable and became known as "The Five Meter Rule."

The air pressure in an inflated habitat can easily support that much mass in the Moon's modest one-sixth g, but it will take years to the dig the deep trenches for habitats for this settlement, especially if low-mass machines were all you can get to the site. Where would the money and workers come from?

The answer was typically twenty-first century: organize a million people over the Internet and just do it. And so the MOVE concept was born. The big data from lunar satellites was reduced, a site was chosen, small exploration rovers were sent, the settlement was designed, the build site was laid out, and, only then, the digging was begun.

This all took time, and there were many setbacks, but the work somehow was done. Someday soon, the players all hoped and somehow knew; a lunar habitat would be flown in from Earth and would crawl, caterpillar fashion, down the ramp onto the flat bottom of their trench.

Then the digger teams could get back to work to cover the habitat with the meters of lunar regolith needed for radiation and thermal protection. It had been a big dig to get the project this far; it would be a big dig to continue. It will always be a never-ending big dig as long as there are people who want to live on the Moon.

The boulder that is their focus today is enormous from a digger's point of view, too big to lift or drag, even with two rovers and even in the Moon's low one-sixth g. There was no option but to blast this obstacle.

There is nothing the MOVE teams love more than a good blast, so it is party time on the virtual Moon. Blasting provided a much-needed relief from the endless drudgery of trenching.

The specially equipped mineral exploration rover, Powderman01, is now finishing drilling a pattern of holes in the boulder and is about to place the explosive charges. The only safe thing to do now is to blow the rock to smithereens.

On Earth, on Skype

"Wizard, did you know Amazon Warrior was leaving?" demanded Princess of Mars. She had opened a private line to him without including the rest of the team.

"I only found out about it a few minutes ago," he assured her.

"She can't do that, can she? I mean, doesn't she have to give notice or something?"

"No, she doesn't have to. We are all volunteers here," the Wizard replied. "Any of us can leave whenever we need to. Of course, this abrupt announcement will have consequences. It's unlikely she will ever get a slot on another lunar MOVE team, not trenching, and certainly not exploring."

"I must talk her out of leaving then," Princess of Mars interjected.

"Too late," said the Wizard, "Just an announcement about possibly leaving sends you to the back of the queue. With thousands of qualified people suddenly in front of you for every possible position, there is nothing left to do but leave. What's done is already done."

"After all her work, she won't have anything more to do with settling the Moon?" Princess of Mars was clearly seeing herself in the Amazon's place one day too soon.

"She has a new role. Now she's a Moon Dig alumnus, and having as many of them as we can in the real world is more important than any single MOVE team member."

"I thought we were doing real things to build a real Moon settlement." Princess of Mars spoke ruefully, in a calmer but sadder voice.

"Yes, we are," continued the Wizard, "and the people of Earth could go ahead with an ambitious human space program any time they get up the political will to do so. The trenches we dig and the rover path-finding expeditions we complete are vital, but our real strength is building up a cohort of hundreds of thousands of people on Earth who have a deep personal commitment to space exploration."

"You make all the hard work we do sound inconsequential," retorted Princess of Mars.

"Our MOVE teams have dug a few tens of meters of trenches and blazed a few hundred kilometers of new trails. The same things could have been done by a big government program in a few years, but the political will was not there. So we've done it as a Massive Online Vetted Exploration. Hundreds of thousands of people on the Web have spent millions of hours studying, training, playing simulation games, and talking it all up. The old 20th century way to explore space got us Apollo to the Moon, but it did not get us back to the Moon in 50 years. So we built a new 21st century process for space exploration in only five."

"And Amazon Warrior was in the MOVE team almost from the start," Princess of Mars emphasized.

"Truth be told, she's been in longer than me," admitted the Wizard. "And now she has moved on."

"Do you know where she is going?"

"Our Amazon Warrior has gotten herself a good position with a high-tech firm. That's just where we need our graduating people to be, in places of power and influence where they can turn our mass spacer movement into real people on the real Moon."

"I still think she should have told us earlier," lamented Princess of Mars.

"Welcome to the 21st century." The Wizard just shrugged and rang off.

On the Virtual Moon

The view of the soon-to-be-smashed boulder was very good, much better than on the real Moon. The computers of the Virtual Moon could combine many views and light settings to build a full panoramic view of a scene and then embellish it.

Moreover, the virtual people could get right up close too. No one here was worried about being hit by a fragment of flying rock.

On the Real Moon

The boulder in question was about the size of a pickup truck and had a distinctive pink vein. On the Moon, in a world of grays with all the hue variation of an old asphalt parking lot, color attracts attention.

Powderman01 had now cleaned out all the drill holes with a puff of gas. It very carefully removed a stick of explosive from a box on its back and slipped on a detonator cap trailing a wire. Then it placed the explosive stick in the first hole and pushed the charge in with a long rod built into its arm.

From this time on, a solar storm could easily provide enough stray electrical charge to blow a stick. Not only would that do in Powderman01, but also the mistimed explosion could not even be counted on to split the boulder properly.

But the Sun is monitored carefully and you can depend on at least a few hours warning for every one of the zero to five deadly solar storms that hit any given site on the Moon each year. When the dig was finished, one of its key functions would be to shield people from just such storms.

Powderman01 finished setting the explosive charges and slowly rolled down the length of the trench, paying out a fine cable. It then huddled down in a small crater just off the trench ramp — it was ready —.

On the Virtual Moon

The huge countdown clock hanging over the trench rolled over to zero and above it a banner made of fireworks appeared reading "Fire in the Hole" and instantly starting popping off flashes. The crowd went wild.

The boulder shuddered, kicked up some dust, and split into pieces. A few small bits of rock went flying in high arcs in the low gravity. Each piece of flying rock was tracked and marked by a spherical image the size of a soccer ball.

The front row of spectators took off at a dead run. Jostling and pushing, they tried to head butt the balls, but to no avail. Jack-of-all stood his ground and aimed his longbow carefully trying to take out as many balls as he could. In the end, he was credited with two hits.

Those remaining in the stands did the wave, a wave that did not simply travel from one end of the stands to the other, but rather one that spread out in complex patterns. Their wave was a stone thrown into still water.

On the Real Moon

The explosion itself did not amount to much. After all, it was designed not to throw splinters and the MOVE teams had voted against big explosive sound effects as simply being too phony in the vacuum of space. A few dozen lines of geophones scattered across the Moon captured the blast.

Powderman01 reentered the trench and surveyed its work. The boulder had been well split and no unexploded charges remained. No chunk of the rock was too large for a two-rover drag out of the trench. The trench's sloped walls had sagged a little in places but nothing the diggers could not handle. Powderman01 carefully picked out samples of the pink vein for further study.

On the Virtual Moon

"I am sorry I could not tell you sooner," apologized the Amazon Warrior. "I did not find out that I really had the job until a few hours ago. You just don't want to spoil something good like a job offer by bragging about it before you truly have it nailed down."

"The Wizard says you can't take your resignation back now and there is nothing we could do now to talk you out of it anyway," said the Princess of Mars, "but I still have a right to be upset."

"You certainly do," conceded the Amazon Warrior. "I wish I could tell you more, but it is all hush-hush. The job is with a new high-tech startup, but I cannot say which because they will be doing an IPO this year. Everything they do now has to be secret just so they cannot be accused of artificially jacking up their initial stock price."

"I hope you get some stock options," added Jack-of-all.

"Yes, I do, but everything turns on the company having a bang-up year," continued the Amazon Warrior. "From now on I will have to put every hour of my time and every ounce of my personal energy into my new company."

"So no more Moon Dig," ventured Jack-of-all. "I hope they gave you credit for all your Moon Dig work."

"Yes, they certainly did," she stated emphatically.

"I would have never gotten this job offer without my Moon Dig certificates. My BA in history certainly does not impress anybody these days."

"Good news!" the Wizard exclaimed. "I just got word from the Moon Dig coordinating committee. Our team will not be disbanded. Amazon Warrior's leaving was ruled normal turnover since she had been in this post for nearly a year. Jack-of-all is our new team commander and a replacement team member will be joining us shortly."

"You mean the whole team could have been disbanded?" interjected the Princess of Mars.

"Yes. With thousands of qualified people ready to take on any position that becomes open and new teams for new rovers so rarely needed, any disruption is likely to fold a team," the Wizard explained. "It takes time for a team to recover after losing a member, so efficiency suffers, but it takes time for a new team to become efficient too.

The decision could have gone either way.

"But we are all still in," said Digger03.

"You will always be in, Digger03," chastised the Princess of Mars. "You are stuck on the Moon till your wheel bearings freeze."

"Do we get to keep our cairn?" asked the Amazon Warrior.

An officially disallowed cairn would very quickly become the source of good rocks for other teams and would simply melt away from the top down.

"Yes, we keep our cairn," the Wizard reassured her.

"We can add the new team member's name to it later."

On the Real Moon

Powderman01 rolled along the trench, picking up a few samples of pink rock and recoiling its detonation cable. Digger02 rolled in as it left to start the cleanup process; Digger03 would follow in turn. The slow pace and grueling work of remotely managed trenching on the Moon picked up right where it had left off. On that quiet mountain ridge, the serenity of the real Moon was now restored.

On the Virtual Moon

The spectators poured from the stands. A swarm of assorted TORDIS's, Flash Gordon spaceships, and fighter jets went whizzing away in the direction of Earth.

The bulk of the crowd jostled around to a draw on the backside of the rise to claim their steeds. A thundering herd soon broke from behind the rise. It was led by the fast-running Cheetahman, who was followed close behind by every conceivable animal, real or imaginary, that humans have ever dreamed of riding. A bunch of skeletal cowboys even drove a herd of glossy black steers with steel hooves and snorts of fire and smoke, such was the Ghost Rider team. The empirical walkers brought up the rear as this screaming pack took a few victory laps around the settlement site. Yet none of them left as much as a single footprint in the endlessly patient blanket of lunar regolith.

The Amazon Warrior slowly turned away from the mesmerized Rocky Horror team, walked a few paces, and simply faded away.

A medieval lady of significant personal bearing, riding a green triceratops, broke from the circling throng. She pulled up the enormous animal in front of the Rocky Horror team and dismounted into a curtsy with a grand sweep of her hand.

"Greetings to my new team," she called out.

"You may call me Guinevere."

DIDACTIC ACTIVITIES

The starting point for a didactically well-structured work is the strategy-programming and the work-planning.

Programming means first of all defining objectives and choosing appropriate teaching methods and techniques to achieve them. The objectives are important because, if they are not well defined, the indispensable basis for the choice and programming of the methodologies, the material, and the didactic contents is missing. It lets the teacher lead students to become aware of their own learning path, to understand its meaning, and to keep alive the motivation for learning.

The approach to language will be communicative and at the same time playful, will start from mainly oral activities that involve the pupil in the first person and put him at the center of the learning process (Cameron, 2001)-

Key Theme

- *"I thought we were doing real things to build a real Moon settlement." Princess of Mars spoke ruefully, in a calmer but sadder voice. "Yes, we are," continued the Wizard, "and the people of Earth could go ahead with an ambitious human space program any time they get up the political will to do so. The trenches we dig and the rover path-finding expeditions we complete are vital, but our real strength is building up a cohort of hundreds of thousands of people on Earth who have a deep personal commitment to space exploration."*

- *"Our MOVE teams have dug a few tens of meters of trenches and blazed a few hundred kilometers of new trails. The same things could have been done by a big government program in a few years, but the political will was not there. So we've done it as a Massive Online Vetted Exploration. Hundreds of thousands of people on the Web have spent millions of hours studying, training, playing simulation games, and talking it all up. The old 20th century way to explore space got us Apollo to the Moon, but it did not get us back to the Moon in 50 years. So we built a new 21st century process for space exploration in only five."*

Pre-Teaching: It is a pre-reading moment useful for the introduction of ideas and words key present in the story, the verification of the vocabulary already known to the students and useful for understanding the text.

Reading: Descriptions or (if added) images play a key role. At first reading it can focus on the screenplay, postponing it to a second reading the analysis and understanding of the message. For the students, it certainly proves very engaging for them to actively participate in reading by miming some scenes. In this way, the understanding of meaning becomes more immediate and less memorization challenging.

Vocabulary Related to Environments (Examples)

- *"The Sun is well up now, a week into the lunar month, but only a few degrees above the horizon."*
- *"The Earth hangs in the black sky due north with its striking whites, greens, browns, and blues."*
- *"The Earth's disk appears much bigger than the lunar disk appears back home, almost four times larger."*
- *"the Sun is due east, and the waning gibbous Earth gives a fine glow, throwing light at right angles to the Sun. "*
- *"That Earth shine lets the hub's cameras peer into many of the otherwise pitch black shadows."*
- *"direct sunshine hits hour after hour, the surface temperature can reach 200 C. "*
- *"in the cold blackness of space, the temperature can drop to -120 C. "*
- *"the regolith carries heat only very poorly so the temperature under as little as one meter of the covering of crushed rock is around 20 C and good for human habitation."*

- *"Despite the workable construction area here being both tens of kilometers long and wide, there are no truly flat spaces."*
- *"Everywhere the surface slopes a little this way and that. The great flat areas of Earth, the planes and ocean abysses, were all created by liquid water. There has never been any liquid water on the Moon, ever."*
- *"Several billion years ago this place was a great mountain range, part of a ring of an enormous basin, the South Pole – Aitken basin."*
- *"Now every feature is worn into rounded shapes by so many eons of meteorite strikes."*
- *"Even the south edge of the massif is not a true cliff but rather simply slopes more and more until the wheels of a rover can no longer get traction in the powdered surface material and the vehicle will simply slide to oblivion into the cold and the dark".*
- *"The entire area is pockmarked with craters ranging in size from pot holes to a house. None are especially deep but they turn the path for a wheeled vehicle into an obstacle course."*
- *"Across this landscape are scattered many rocks of all sizes, from pebbles to boulders. These erratics were blasted here by distant great meteor strikes and provide samples of the lunar crust from many locations and many ages."*
- *"The Scott A Massif is a plateau running east to west and standing about six kilometers high. It has local high points with sunlight for power about 95% of the time as its southern edge is a peak of eternal light, and it also has continuous line-of-sight to Earth. It has a kilometers-wide buildable area and at least four routes that never exceed a 20% grade running a long way down to the permanently shadowed craters far below and to the south."*

Technology Related Vocabulary (Examples)

- *"three racks of state-of-the-art electronic equipment"*
- *"dozens of blades with the latest deep learning chips"*
- *"virtual reality hardware"*
- *"power, cooling, and row-upon-row of routers to handle the heavy incoming traffic."*
- *"advanced security firewall to protect the entire system"*
- *"on-line courses"*

- *"a remote robotics operator."*
- *"a communication hub"*
- *"a billboard sized photovoltaic panel"*
- *"set of cameras"* .
- *"the solar powered excavator"*
- *"two small telescopes"*
- *"laser communication*
- *"powder-fine grit"*
- *"cairns"*
- *"bit of gold blanket or a scrap of metal",*
- *"The radiation"*
- *"Earth's atmosphere"* .
- *"The big data from lunar satellites"*

Presentation of the Characters: The exercise is impersonating characters (answering questions: What's your name, How old are you ? Where are you from? What are you wearing?), trying to act and interact as the character would have done. The dramatization proves to be a positive and productive activity. Moreover it encourages the student to express himself without inhibitions and allows him to customize a text.

Examples: In the real Earth *"The Wizard is a senior IT professional; the Amazon Warrior is a recent liberal arts graduate; the Princess of Mars is a perpetual student; the denizen of Sherwood Forest is a martial arts instructor".*

Scenes: The teacher can ask the students to represent the different scenes of the story and write alongside them keywords. The story can be recited and recorded by creating a movie, writing an article for the school newspaper and contributing to the creation of the school website.

Retelling the Story: Students attention is focused entirely on the meaning of the story. If understood, they can be encouraged to change some actions or the even the ending of the story.

Skills Involved

- *Comprehension:* Understanding the story; listening to and identifying key words related to interactions between characters
- *Interaction:* Participate in a recited version of a story through mime and repetition of words

- *Image Education:* Graphically representing scenes from the story, using different techniques; creation of a comic book of personal reworking of the story
- *Citizenship Education:* Collaborating within the class group, in a small group and in pairs; consolidating the identity of the class group; reflecting on scientific issues and the technical development of the society. Examples of social issues:
    - *With thousands of qualified people ready to take on any position that becomes open and new teams for new rovers so rarely needed, any disruption is likely to fold a team," the Wizard explained.*
    - *"It takes time for a team to recover after losing a member, so efficiency suffers, but it takes time for a new team to become efficient too.*
    - *Our Amazon Warrior has gotten herself a good position with a high-tech firm. That's just where we need our graduating people to be, in places of power and influence where they can turn our mass spacer movement into real people on the real Moon."*

TECHNICAL DISCUSSION

New Approach to Space Exploration

This short story, "The Big Moon Dig", proposes a new approach to human space exploration. It is a proposal to do many of the things that are already hot and keep doing them until a way is found to make some of the heat rub off on human space exploration itself.

This approach is here called Massive On-line Vetted Expedition (MOVE) and is a plan to execute one grassroots move after another until human beings are back on the Moon. MOVE (Riley, 2017a) starts with on-line games [Hot] based on real science data [Hot to some].

To play the game a player will need technical training in real tele-robotic operation and space settlement design though a series of Massive On-line Open Course (MOOC) [Hot]. To improve the game, everything will be crowd designed [Hot] and tested so the player must spend the hours needed to understand the mass of available data on the Moon.

If the training and game catch-on, that is go viral, then a large and enthusiastic fan base will be generated for the next step, a small exploration rover on the Moon. Such a delivery is commercially available today for around 1.5 million dollars a kilo -- the Moon is metric --. Such a price is within the reach of successful games today. The big prize in the early game then is to become a tele-operator of a real lunar rover.

If the viral energy continues, then the resources to ship small exploration rovers could be found. Then a digger rover could follow to start construction of the first real lunar settlement, here called Armstrongville, Moon. This construction period is the time of this series.

Only when all is ready and an enormous number of people are highly trained for every task, will the actual habitats be shipped. Then and only then, people can return to the Moon, safely, permanently.

A review of recent space exploration efforts reveled two problems blocking space settlement in the 21st century:

1. **Radiation Protection:** In the Year 2000 two space instruments were flown that were designed to specifically identify the long-term danger of radiation to humans in space. (CRaTER) (MARIE) The answer came back that the problem was twice as bad as previously assumed and that earlier designs had not provided adequate protection even at the lower assumed levels.
2. **No Funding Prospects:** Apollo to the Moon was financed by the government primarily because of the Cold War. The Cold War is history. Currently there are no prospects for funding for a large government financed space initiative.

There are however a number of strengths that were not available at the time of Apollo to the Moon:

1. **The Internet and Social Media:** It is now possible to organize millions of people with a common interest to be in action on a major project.
2. **Commercially Developed Space Hardware:** Commercial businesses are building and testing major Earth to Low Earth Orbit (LEO) delivery systems. These operate at significantly reduced cost compared to earlier systems, like the Shuttle, and are a major new asset.
3. **National Space hardware, Facilities, and Data:** The United States and other governments continue to maintain major launch sites and launch scientific satellites provide new data.
4. **Computers and Robotics:** The entire field of computers and robotics is exploding. AI will soon get a bit boost with the commercial availability

Figure 6. Deep Habitat (original design by T. Riley)

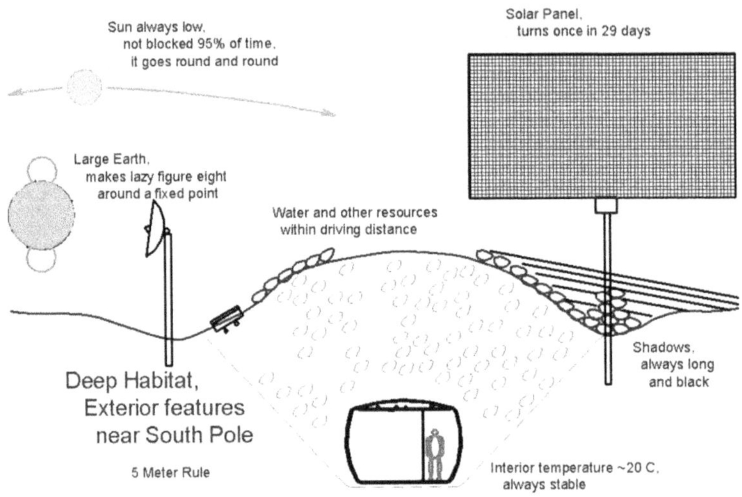

of chip sets specifically for AI. They will then be fast, cheap, and low powered.

5. **Peaks of Eternal Light (key term):** These are locations near the lunar poles where Sun light is available nearly all the time.
6. **Permanently Shadowed Craters (key term):** Lunar craters near the poles whose bottoms have not seen Sun light in billions of years and are therefore very cold. Volatiles like water have accumulated in these locations over the eons.

Discussion of the problems and the new tools lead to the following basic ideas that could allow the lunar settlement effort to move forward:

1. **Lunar Regolith as Shielding:** The amount of regolith (key term) needed to provide habitat shield is large but not so large as to be unworkable.
2. **Mass Movement to Provide Manpower:** A mass movement, MOVE (key term) to return to space could be organized using the latest social media.

Concept development work for this approach was started in 2012 and produced:

1. **Big Moon Dig Web Site Established:** (Google "Big Moon Dig")

2. **Premium Lunar Settlement Site Identified:** The Scott A Massif site (key term) for a lunar settlement.

By 2015, the Big Moon Dig was receiving the same criticisms that Apollo to the Moon met. If society could pay to go to the Moon, why does society not work directly on the problems of people on Earth?

To address these criticisms the Big Moon Dig started work on using the MOVE concept to address many down to Earth problems. It became clear that this would require meaningful communication with a massive number of young people which, in turn lead to this book.

Why Dig?

Human beings evolved in an environment with only modest background radiation thanks to the Earth's safe environment. Unfortunately, the Space environment includes a substantial amount of radiation beyond what human bodies can tolerate. For human beings to live for long periods of time in Space, they will need sufficient protection from radiation at least to match the Earth environment in which they evolved.

Early in the 21st century, NASA flew two missions with instruments that actually measured the damage expected to human tissue for the first time. These were the Lunar Reconnaissance Orbiter (LRO) instrument, Cosmic Ray Telescope for the Effects of Radiation (CRaTER), and a similar Mars mission instrument, Mars Radiation Environment Experiment (MARIE). These instruments used a plastic block that reacted to radiation like human flesh. The results were very upsetting. The damage was about twice the level that had been previously assumed.

This result means that virtually every space station and manned ship that had been planned to date was simply a death trap, at least in the long-term. Radical design changes are needed.

There are two major types of radiation and protection must cover both:

1. **Galactic Cosmic Rays (GCR):** This radiation comes from far outside the solar system. It contains highly energetic particles that can penetrate deeply (see Fig. 9).
2. **Coronal Mass Ejections (CME):** These are large masses of elementary particle, comparable in mass to a mountain, which are occasionally ejected by magnetic storms on the Sun. The particles do not penetrate well but there is enough of them to cause serious health consequences to

Figure 7. Earth's particle environment (dominated by galactic cosmic rays and solar particles) and the two main radiation (Van Allen) belts around the Earth (Source: https://www.researchgate.net/publication/318029811_Space_as_a_Tool_for_Astrobiology_Review_and_Recommendations_for_Experimentations_in_Earth_Orbit_and_Beyond/figures?lo=1))

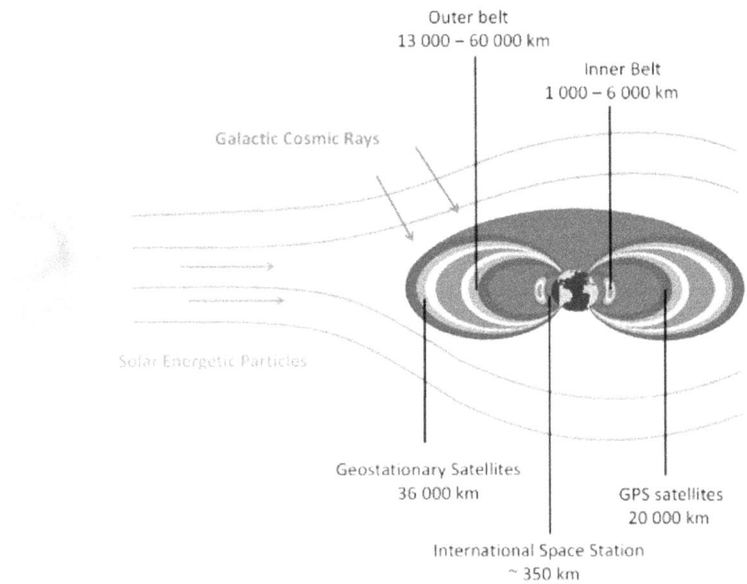

Figure 8. Stellar storms (Source: https://futurism.com/stellar-storms-solar-flares-coronal-mass-ejections)

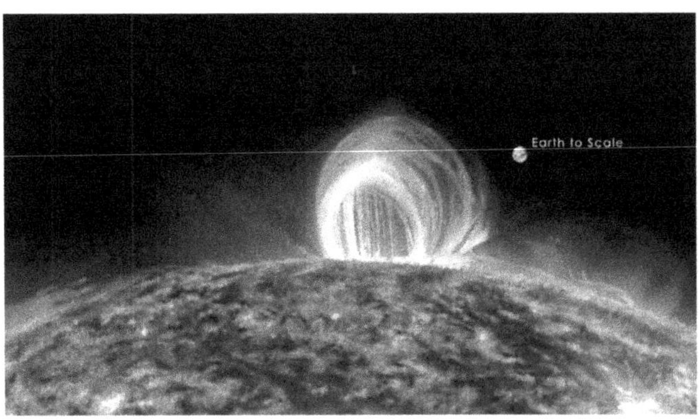

a human if exposed with only the protection of a spacesuit or spaceship skin. Any location on the Moon can expect zero to five such storms every year that last from 24 to 120 hours. (see Fig. 7 and 8).

The problem for settlement design then is to stop both types of radiation at least as well as they are stopped on Earth.

Many recent designs call for an additive shell covered with fused regolith. This produces walls one to two meters in thickness. Such walls stop the radiation from CEM's acceptably well but the deeply penetrating radiation in GCRs is not stopped but rather converted into cascades of weaker particles. Two meters of regolith is simply not enough shielding and it even makes the GCR radiation worse on human tissue.

In these stories, the approach to radiation protection chosen is a very thick layer of lunar regolith. This minimizes the amount of mass that must be shipped from Earth and thereby keeps the cost of the project under control.

The Big Moon Dig Project (Riley, 2017) calculated the thickness of regolith needed to match the mass of the Earth atmosphere at between four and five meters. Providing this level of protection is a big challenge; major excavations on the Moon are required. Fortunately, farther calculations indicated that an inflatable habitat could easily hold up this mass.

Why Not Utilize Lava Tubes on the Moon?

There is another possibility for obtaining the critical radiation shielding that the lunar settlers will need, Lava Tubes and their Skylights (Pettit, 2018).

On Earth, lava commonly flows from a volcano in rivulets. These sometimes cool on the surface forming a solid crust. If the lava, now insulated by that crust, continues to flow away underneath, a cave like structure can form called a lava tube. Often the roof of lava tube caves in forming an opening to the outside, a skylight.

There was a time in the Moon's early history when lava flowed frequently there. Now there is clear photographic evidence that very large lava tubes commonly formed but later, unfortunately, their roofs caved in completely.

There are also some very deep holes that are clearly not meteor craters. These appear to be skylights for more deeply buried lava tubes. For both good and bad, lunar regolith contains one to two percent metallic particles. These are the result of iron and stony-iron meteorites hitting the Moon. The metal component vaporizes into a metallic steam that then cools on the fly to form minute spheres.

The spheres composed mostly of iron and nickel are ferromagnetic and adsorb the magnetic component of radio waves well. The result is that radar cannot penetrate lunar regolith far enough to detect lava tubes so no one has been able to map them properly.

Recently the GRAIL mission (GRAIL) (Williams, 2016) has been able to map the Moon's complex gravitational field in enough detail to show missing mass in several areas where lava tubes are suspected. Unfortunately, these maps are quite course and are of limited use for planning a settlement.

How much room is available in a lava tube? What are the floors like? Are the floors covered with rubble? How stable are the skylight edges? The answers to these questions are just not known and practical space missions cannot be counted on to find out under available budgets.

The lava tubes remain impractical for lunar settlement planning especially for the principle task described in these stories, that of getting humans back to the Moon soon. For the next effort, a fast and cheap approach is needed. To use lava tubes the mission would either have to land directly down the skylight or build a landing facility on the edge of the skylight and lower materials into it.

The first approach is dangerous and has problems with both communication with Earth and sunlight for power. The second approach requires building a major receiving installation that would have only limited radiation protection, and that must be built before the actual settlement is even started.

Although the approach of digging trenches that is used for these stories is, at best, odorous, and the lava tubes promise future cities on the Moon not just small settlements, the next practical step is to dig and thereby get back into the game. Only much later can humans expand from settlements to cities and in the process once again become cave dwellers.

CONCLUSION AND FUTURE RESEARCH DIRECTIONS

What Is Hot and What Is Not?

Right now human space exploration is simply not hot. Can it be reorganized to take advantage of the hot, powerful innovations of the 21st century?

Some people are working on the commercialization of space and that is fine but not enough by itself. Commercialization will work well for efforts that have clear monetary value, like communication satellites, but it will do nothing for general space exploration, which is much more likely to generate value to society rather than cold hard cash.

To make matters worse, recent robotic space expeditions (CRaTER) (MARIE) have shown that the danger from radiation for people in space is about twice what people had thought. This means that all of the spaceships and space stations that have ever been shown in movies and TV are just plain deathtraps. People will need to redesign them all.

So what is hot right now? Electronic games are hot. MOOC's, (Massively On-line Open Courses), are hot. Robots are hot. AIs are hot. Crowd source efforts are hot. Yet, human space exploration has **not** been hot for several decades now. Maybe it is time to try a different approach.

Transition

This first story in this series is about new approaches to human space exploration. It showcases the intense fulfillment many humans can have in participating in a great common effort. It is largely about people interacting with people.

In our next story, "Stanford's Digs", the focus turns to the relationship between humans and their machines. This story looks at problems in building a mutually beneficial relationship, referred to in these stories as the Human/ Machine Symbioses.

REFERENCES

Cameron, L. (2001). *Teaching Languages to Young Learners*. Cambridge University Press.

Fontana, A. (2009). *Manuale di storytelling. Raccontare con efficacia prodotti, marchi e identità d'impresa*. Rizzoli Ed. doi:10.1017/CBO9780511733109

Kampler, A. (2010). *Obama's 'Game-changing' NASA Plan Folds Constellation, Bets Commercial*. Retrieved from http://spacenews.com/obamas-game-changing-nasa%E2%80%82plan-folds-constellation-bets-commercial/

Riley, T. (2017). *The Big Moon Dig*. Retrieved from http://bigmoondig.com/BigMoonDig.html

Volpicelli, G. (2015). *Nasa releases plans for the Orion space capsule's first mission as it prepares to head to Mars*. Retrieved from http://www.dailymail.co.uk/sciencetech/article-3356404/To-Moon-Nasa-releases-plans-Orion-space-capsule-s-mission-prepares-head-Mars.html

Williams, M. (2016). *GRAIL Data Points To Possible Lava Tubes On The Moon*. Retrieved from https://www.universetoday.com/128208/buried-lave-tubes/

KEY TERMS AND DEFINITIONS

Apollo to the Moon: The formal name of the major space program run by the United States that put human beings on the Moon in 1969. It provided a positive vision of an expanding future that the Big Moon Dig is trying to reestablish.

Big Moon Dig (BMD): A fictional grassroots space movement with the purpose of building a settlement on the Moon. It is a MOVE organization. Historically Apollo to the Moon provided an enormous number of people with a vivid, positive vision of the future. Such a vision would be a major asset in addressing the great problems of the 21st century if it can be reestablished. The Big Moon Dig's top-level purpose then is to rebuild that vision.

Corporate Person: In the United States, a corporation has some, but not all, of the rights and responsibilities of a person. The major AIs in these stories are there for independently incorporated to make them some kind of a person and to provide human supervision through their board of directors.

Digger03: This is a lunar excavator, corporate person, and AI member of the Rocky Horror Team. It is specifically designed to work in teams with humans.

Interface Avatar: A representation used by a person, real or AI, which can interact with other avatars in a virtual environment. Their appearance is a personal choice of the person and include many fictional characters. In these stories, where there are multiple version of the same character they are identified with different colors. The avatar is isolated from the main AI so that if it is damaged the main AI remains safe.

Massive Online Vetted Exploration (MOVE): This is a fictional means of organizing millions of people over the internet to take on specific exploration missions such as building a settlement on the Moon.

Peak of Eternal Light: There are high points near the poles of the Moon that receive sunlight nearly all the time. These are simply accidents of geology. They are attractive sites for a human settlement as they provide an answer to the problems of powering the settlement through the 14-Earth-day-long lunar night.

Permanently Shadowed Crater: Any of several meteor craters near the lunar poles that have not seen the heat of the sun in more than a billion years. Because of their extreme cold, they accumulate volatile materials like water. They hold great promise as sources of materials for a human settlement but will be very difficult to mine due to the cryogenic cold and lack of sunlight for power.

Rack: A metal cabinet specifically designed for electronic equipment. Originally called a Relay Rack and designed before 1900 by the Bell Telephone Company. The original relay racks were 19 inches between the supports and had a complex bolt pattern. This is an example of a long-standing convention in design that will be extremely difficult to convert to the metric system. In these stories most of the electronics of the AI's is assumed to be mounted in such racks.

Regolith: The finely pulverized rock that covers most airless bodies. It is the result of billions of years of gardening by micrometeorites. It does not qualify as soil as it contains no water or organic material. Lunar regolith not only contains no water but also contains no minerals formed in water. The top six centimeters are very finely ground and are a major problem for all seals and bearings.

Scott A Massif: A major rock formation near the South Pole of the Moon and the proposed location for a lunar settlement. It runs from 20 to 45 degrees longitude and -84.7 to -86.1 latitude. The Scott A Massif is a remnant of one of the rings of the very old South Pole-Aitken Basin. It is a real plateau about 120 kilometers from the lunar South Pole. Its southern edge is a Peak of Eternal Light and it has at least three trails, that do not exceed a grade of 20 degrees, leading down to several permanently shadowed craters below.

Settlement: The people and structures that allow human beings to live and flourish in a distant and challenging environment. Note that the word "colony" is never used as it has such a negative history on Earth.

Skype: An existing App for communicating between computers featuring both sound and pictures and often used for business communications.

Virtual Moon: In these stories, an enormous virtual reality environment and the primary setting for action. This environment shows the progress made on building the lunar settlement and provides a place for the various interface avatars to interact.

Chapter 2 The Humanity of an Artificial Intelligence

ABSTRACT

Artificial intelligence (AI) is quickly becoming a powerful tool for addressing humanity's great problems of the 21st century. The working relationship between humans and machines then is a key concept in both their futures. If powerfully implemented, this relationship may become a human/machine symbiosis. The institutions that will be needed to establish this symbiosis will require both public consideration and technical design work. The story in this chapter, "Stanford's Digs," looks at the developing relationship between humans and AIs, particularly when working in teams together. The concerns when AI interacts with AI are then considered.

INTRODUCTION

The rhythm of exciting technological breakthroughs is quickening. In recent years, the communications revolution offered an unprecedented access to information and boosted cross-border collaboration.

Today, the big topic is data and, increasingly, the Artificial Intelligence [AI] revolution, which empowers hierarchical organizations to control and sift that information. However, this latest wave of innovation is taking place at a time of broader tensions within the internet and technology spaces that will play an important role for global markets in 2018 (Bremmer & Kupchan, 2018).

DOI: 10.4018/978-1-5225-8401-8.ch002

Table 1. Generation of Computers (Source: Eugene, 2017)

Criteria	Classification
ERAS	• **Pre Mechanical Era** (before 1450) • **Mechanical Era** (1450-1840) • **Elettro Mechanical Era** (1840-1940) • **Elettronic Era** (1940 to date)
ELECTRONIC ERA	• **First Generation** (1940 - 1946) • The Automatic Sequence Calculator was the first automatic computer invented in 1939 • ENIAC (Electronic Numerical Integrator and Calculator) invented in 1946 is considered the first electronic digital computer • EDVAC (Electronic Discrete Variable Automatic Computer) invented in 1948, is considered the first digital computer that could store programs • **Second Generation** (1947 - 1963) • It began with the invention of the transistor in 1947 • **Third Generation** (1964 - 1971) • It began with the invention of Integrated circuits (ICs): computers became small in size, low in price and highly speed. • **Fourth Generation** (1972 - 1985) • Micro Processor was born • MIT built the ALT AIR 8800 computer in 1975. In 1981 the first personal computer IBM-PC was built.. Following this, the Apple computer Company built the Mackintosh Computer in 1984. • Graphical User Interface (GUI) was incorporated • **Fifth Generation (1986 to date)** • Neural networks • Artificial Intelligence • Robot technology • Automatic vehicles • Molecular Computer / DNA computers and Biological Computers • Quantum Computer

There are at least 16 companies who have major Research and Development [R&D] efforts to produce chip sets specifically for AI (Wang, 2017). These promise to make AI cheap, fast, and to require only low power. A wave of computer projects featuring these chip sets in consumer devices is expected in 2019. A significant competition is expected between many of these companies to determine whose chip set will become the industrial standard.

Many computer uses will remember the introduction of the super-graphics chips into the electronic game world starting in the late 1990s. Suddently computer games went from slow with limited graphics often with simple characters jumping up and down, to enormous new worlds in simulated 3D and near movie speed and reality. Just such a revolution is now expected for AI as the new chip sets hit the market.

If society is to keep reasonable control over its AIs, action must be taken soon.

Table 2. Computer typse (Source: Eugene, 2017)

Criteria	Classification
COMPUTATIONAL METHOD, TYPES, TECHNOLOGY	• **Digital Computer:** designed by using the binary digital number system • **Analog Computer:** designed considering the physical data (such as voltage, resistance, temperature, pressure) • **Hybrid Computer**
SIZE AND CAPABILITY	• **Supercomputer:** the largest and fastest computers available but typically used in specialized area as in Defense, Aircraft design, computer generated movies, weather research etc. • **Mainframe Computer:** less powerful and cheaper than supercomputers, very large in size, capable of handling scientific and business applications, such as payroll computations, accounting, business transactions, information retrieval and airline seat reservations • **Mini Computer:** performing data processing activities in a smaller scale. typically supporting not more than 10 to 12 terminals • **Microcomputer:** (Personal Computer systems) consisting of a microprocessor and associated storage and I/O devices, with very small dimensions, having entire circuitry (CPU) in a small silicon chip.

Figure 1. The evolution from a gigaflop to petaflops (source: https://afeatheradrift.files.wordpress.com/2011/06/petaflop.jpg

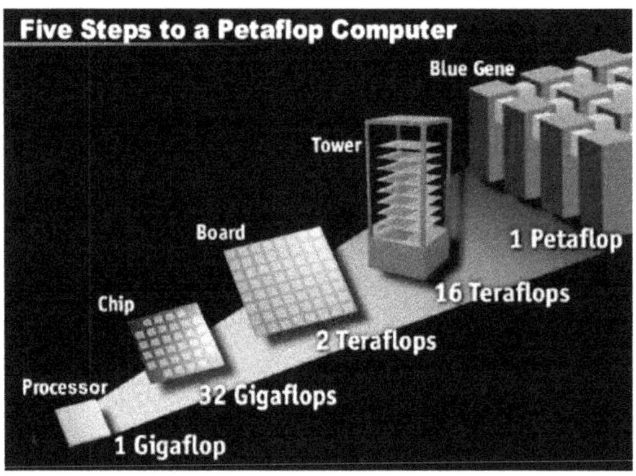

The Great Computer Centers

The most powerful of the new AIs will not be directly in consumer products but will instead be physically located in major computer centers. There it is a straightforward task to have row after row of electronic racks loaded with cards [sometimes called "Blades"] full of the new AI chips in addition to the row after row of routers they maintain to allow millions of customers to directly use this new feature. These computer centers will also provide the power, cooling, and firewalls needed for nearly unrestricted AI operation.

AI Interfacing

A few of the new AI chips sets will be in robots but most people will interact with them as voices in a machine, or as an interface avatar on a screen.

Some major devices, such as self-driving cars and some industrial equipment, will have built-in AIs with voice and graphic interfaces. Such devices have long appeared in popular fiction and people already expect their car to take commands and are quite use to listening to the voice of the radio.

People easily accept talking to a machine and listening to the reply; after all they have been using telephones for more than a century. Boxes you give commands to are now widely available on the commercial market. Many people do object to the box listening all the time, but this problem can be addressed with either a button or very precise first command word.

The graphic Interface Avatar will have the appearance of any thing the users wants, be it people, robots, or trolls. They will move with the person from screen to screen, from cell phone to game console, from TV control to their car.

The appearance and voice of these personal Interface avatars will be personified in a great many forms but they are all subject to a specific rule: the Uncanny Valley. Detailed studies of many customer groups have shown that consumers will accept interfaces that are not very human, perhaps obviously a robot, or interfaces that are very human indeed. Interfaces that try to be human and fail occur to many customers as "sick people" and are consequently rejected.

It is not however necessary that the interface be so good that it is commonly mistaken for a real human being. Many customers do not like to be fooled and are more comfortable if they are in the know.

AI's Teaming With People

There have been many movies where AI robots ran amuck [The Terminator, etc.]. Although this problem is very unlikely, it will simply make the public feel more secure if designers take reasonable steps to prevent it. The approach explored in these stories is to design the AIs to team with humans at every level.

This starts with their basic software design being based on a Human/Machine Symbiosis and not on the now very common Master/Slave. This establishes the concept of everybody helping everybody else.

The larger of the AI in these stories are also independently incorporated. This makes them a corporate person with limited rights and responsibilities. It also puts them under the control of a board of directors most of whom would be human. It remains to be seen if this legal trick leaves the AI as a second class citizen.

For most individuals this teaming will occur as having a virtual friend who does things for them and answers questions. It is also at their side as their wingman in the world of electronic games.

For people in important positions who can tolerate a larger expenditure, their AI will be a key one of their people. The AI will effectively double their personal power by covering such tasks as reducing big data and preparing presentations.

How This Chapter Works

This chapter is focused on Supercomputing, and Artificial Intelligence as a key component of these machines.

A needed premise is that some researchers estimate the chance is about one in three that this development turns out to be ‘bad’ or ‘extremely bad’ for humanity. There is, in some quarters, concern about high–level machine intelligence and superintelligent AI coming up in a few decades, bringing with it significant risks for humanity. They have issued the warning that if the human race ever builds a super computer much smarter than humans are then the human race is doomed to be stuck in a fading existence on a backwater Earth while the super computer grows ever smarter and moves out into the universe (Müller & Bostrom, 2014).

The following Science Fiction story is **not** based on this idea at all. In fact, all these fiction stories based on Artificial Intelligence topics are **not** based the common idea of human and machines as Master/Slave at all.

Quoting M. Tegmark a MIT researcher (Tegmark, 2017): “I think Hollywood has got us worrying about the wrong thing,” he says. “This fear of machines turning conscious and evil is a red herring. The real worry with advanced AI is not malevolence but competence. If you have superintelligent AI, then by definition it’s very good at attaining its goals, but we need to be sure those goals are aligned with ours. I don’t hate ants, but if you put me in charge of building a green-energy hydroelectric plant in an anthill area, too bad for the ants. We don’t want to put ourselves in the position of those ants.” (Andrew, 2018).

The title of this story is "Stanford's Digs", the second one of Big Moon Dig Series. Its analysis focuses on four parts: an introductory didactical analysis, the Science Fiction [SF] story, a proposal of educational activities, and a technical in-depth analysis.

DIDACTIC ANALYSIS OF "STANFORD'S DIGS"

Topics for Discussion

The following discussion points come from information covered in this chapter:

1. How important will Artificial Intelligences [AIs] be to addressing the great problems of the 21st century?
2. What level of control should humans have over the AIs? Is it simply the old Master/Slave, or is there possibly a Human/Machine Symbiosis?
3. What are the rights and responsibilities of a Corporate Person? Is incorporating an AI just a cheap trick?
4. Is there a meaningful difference between human space exploration [using all manner of tools including AIs] and Human/Machine space exploration where all participants work in a symbiosis?
5. If AIs and people work together, who should lead the way?

The story is analyzed as science fiction storytelling, highlighting how it can be used both to explore crucial science issues on the settlement of the Moon and improve computational thinking and creative problem-solving.

What is the difference between the narration of a fantasy world in a wonderful past ["Once upon a time"] or in a timeless world, and the narration of a possible real future, maybe with better life prospects or totally different from actual experience but with human relationships well known to the reader? Both are open to fantasy, but the futuristic ones trigger curiosity and mobilize people to change something. In these stories, the author also introduces real technological elements and the truth of a world that has already developed the ability to shift between the virtual and the real.

In these stories, the Janets are a series of fictional high-end AIs that are specially designed to manage major technical projects intended to address the great problems of the 21st century [global warming, sea level rise, etc.]. As all the units in this series have female interface avatars, they are addressed as female.

The major series of AI are given interface avatars taken from either fiction or history. This reduces the task of the designers who would otherwise have to invent something completely new and makes the interface avatars easier for other people to interface and team. The Janets are based a character from Australian television, Janet King, who was played brilliantly by Marta Dusseldorp. This series has appeared on Public Television in the United States.

JanetM is one of a series of fictional high-end AIs that are managers of major technical projects. The foundation of their design is the Human/Machine Symbiosis and definitely not Master/Slave. JanetM has specific responsibilities in raising funds for all the Non-Governmental Organizations [NGO] efforts to address the great problems of the 21st century.

Environments

The scenes in this series of stories take place in parallel in four environments: on the real Moon, on a virtual Moon, on the real Earth. In this story, much of the action also takes place in another virtual reality controlled by a major computing center at Stanford University, USA that appears to be a secure apartment house for AIs.

Events and Actions

On the Moon, the construction of the lunar settlement is suspended for a few days during the portion of the lunar month where sunlight for power is most limited. This pause is used to upgrade the equipment and software of the AIs.

A major Earth-bound AI, Standford01, uses this pause to develop a strong relationship with a minor lunar-based AI, Digger03. This action disturbs several of the human members of the team.

Main Characters

The main characters of this story are both humans and AIs with the main story line being the development of their working relationship:

1. **Digger03:** This team member is a robot/AI who works as an excavator on the Moon. It appears as a physical device on the Real Moon but also has an interface avatar in the Virtual Moon. It is independently incorporated and so has some, but not all, of the rights of a person. It does not have

a gender and does not represent itself in virtual space as either sex and so is always referred to with the pronoun it. Its lunar-based AI is smart enough to keep the machine out of physical danger on the Moon, but not much more than that. Its electronics on the Moon are supported by a larger rack of electronics back on Earth that enables a full interface. It is a major character in the first three stories in the Big Moon Dig series.

2. **Plum Princes of Mars:** This image on many screens is an interface avatar who is operated by a woman named Clarice. She is a founding member of the Rocky Horror team and a member of Digger03's board of directors.

This character is taken from the classic science fiction books of Edgar Rice Borrows that are set on Mars. This is also the basis of the costume worn by Princess Leia Organa of Alderaain in the Jubba the Hut sequence in the Star Wars franchise. As this is a famous science fictional character many people would like to use it for their avatar so each clone is given a different color to distinguish it, hence the designation and color Plum here. This costume is completely impractical for any type of physical activity, especially fighting, and is simply for show.

3. **Wizard** or **the old man:** This character appears in several of the Big Moon Dig Series. He is a long-term member of the lunar settlement effort and the counselor of the Rocky Horror team and a member of Digger03's board of directors. In real life, he is an IT professional. He often expresses the views of the author of the series.
4. **Stanford01:** This is a major AI developed at Stanford University in California. It is currently limited to Earth but would very much like to get a toehold in space.
5. **Harrison:** A member of Digger03's board of directors who is very skeptical of a major AI like Stanford01 having a seat on a minor AI like Digger03's board.
6. **Clancy:** A member of Digger03's board of directors who is very good at playing games.

Most of these characters form the Rocky Horror Team that operates Digger03 on the day shift or are members of Digger03's board of directors, or both. On Earth, they are a few of the scattered volunteers in the MOVE project of the Big Moon Dig.

Technology

The main cited technologies are: Virtual Reality and Artificial Intelligence applications. For a specific in-deph discussion, see the dedicated Technology Section below which includes supporting topics like design considerations for AIs.

Communicative Approach

There is no doubt that in the communication process the oral language is of primary importance, as well as the starting point for learning a second language.

Stories motivate learning; they are fun and help develop a positive attitude especially towards the second language. The children are involved personally because they often identify with the characters and try to interpret the story and the illustrations thus favoring the development of creativity.

- Listening to stories helps children pay attention to the rhythm, intonation, and pronunciation.
- Storytelling facilitates the development of learning strategies such as listening to grasp the general sense, predicting later scenes, guessing meanings, and hypothesizing.
- This story allows the teacher to introduce and revise words and structures, exposing the learners to the language and enriching their linguistic heritage, especially technical keywords.
- In the case of a foreign language, the narration offers opportunities to present cultural information of the country in which the language is spoken. Every word has its own history, its particular meaning, its value that is not always easily translated into another language but that with the right material the teacher can try to “pass” to the students, immersing them in the atmosphere of the second language.

Playful Approach

It is therefore important to cut out moments dedicated to activities within the class substantive games and activities, capable of producing true emotional involvement, a fundamental engine of any form of learning.

Figure 2. Digger03 on the Real Moon (Source: Riley's Image)

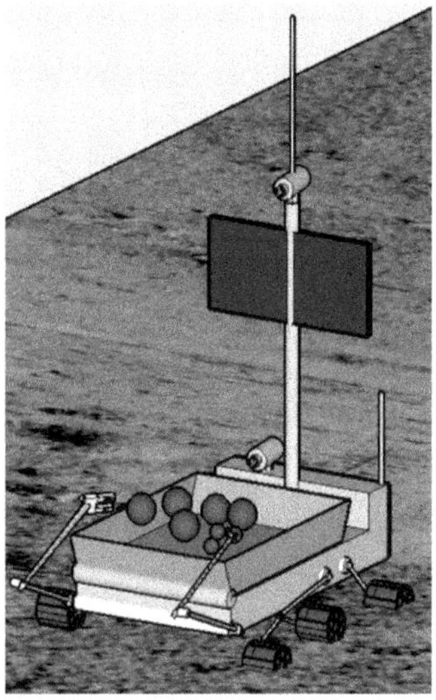

Transition Into the Story

In the Big Moon Dig stories, people want so much to participate in human space exploration that they will devote years of their lives to training and then days to the boring tasks of operating a lunar rover. These people clearly have purpose.

If a person tells an AI to do something, the AI then has a task and with it a purpose too. However, can an AI take on a purpose all by itself?

If that new purpose of the AI were space exploration, would people object? What manner of tricks would the AI have to use to fulfill its new purpose? Is there a meaningful difference between human space exploration [using all manner of tools including AIs] and Human/Machine space exploration where humans and machines all work together in a symbiosis?

THE FICTION STORY: "STANFORD'S DIGS" BY TOM RILEY

Premise

"In Space, sometimes you need a friend, even if you are a robot – especially if you are a robot –."

On the Real Moon

Twenty rocks each the size of a melon raddled out of Digger03's dump bed into a pile. There was little sound in the hard vacuum only the shaking of the robot's own body. It pushed the rocks around with its long, thin arms until it was happy with its work.

"I will take a power break now," said Digger03. "I need 18 minutes in this light patch before I can safely pick up my next load."

Clarice surveyed the scene through Digger03's eyes. The Sun was low and way to the south. Next week the Sun would take its monthly trip though the distant mountains. The works site had power for now but a long rest period was coming up. Clarice for one could use a rest.

The lunar surface was gray, endless gray. Where the sunlight hit it was blindingly bright. Where there was shade, there was true shade – black shapes of fathomless depth –. With the sun angle now so low, every rock and rise threw black lines across the steel gray-white terrain. Magnificent desolation it has been called and magnificent desolation it is.

The only evidence of man at this future lunar settlement site, located on the Scott A Massif, was the half completed trenches and piles of rocks call cairns that served as survey markers. Scratched on a flat rock deep inside one of these cairns, written in Digger03's uneven scrawl with a diamond stylus, was Clarice's name listed among those in her early digger team. A rock-of-names can last for a million years on the Moon.

Digger03 stood in a trench running east to west that was now about 16 meters wide, 30 meters long, and, now unfinished, only 1.2 meters deep. The Big Moon Dig was well begun but still had a very long way to go. At this time of the lunar month, Digger03 had to get its power by moving from patch of light to patch of light with its solar panels held way up high.

To the north hung the Earth. It was low in the sky but clearly above the horizon, an easy target for the station's directional antennas. Its yearly path was a vertical figure eight. The daytime disk was big and a bright mix of blue, and green, and white, and brown.

Inc. Digger Falcon Zero Three was a robotic excavation machine especially built for the Moon. It's hardware was built light but very strong. Its solar power panel stuck straight up on a pole like a protest sign that could spin around to find the low-angled Sun in locations near the lunar poles. Its brains were divided between the Moon and the Earth, which gave him a bit of a split personality, and the 1.3-second transmission delay really slowed what thinking it could do. It had all the brains it needed to do its job, even if they were a little out of date. And, of course, Digger03 was independently incorporated so it was a corporate person, legally recognized as such by none other than the United States Supreme Court.

Digger03 was slowly, precisely digging a great big trench on the Moon, aided and abetted by the Earth-bound Digger03 Shift 1 Team, the Rocky Horror, which was Clarice's long-term team. In a few years that trench would protect an inflated human habitat in a place that would then be called Armstrongville, the Scott A Massif, Moon.

Someday habitat sections will be flown from Earth. They will arrive in a very compact form later to be inflated. They will then move in the fashion of a caterpillar across the surface and into the trench, there to be interconnected.

Once the fine broken rock, or regolith, was moved again to cover over the inflated habitats, the filled trenches would provide the protection people need from radiation, from thermal extremes, and from micrometeorites. It was free protection, at least, free for the moving of it. Many great trenches will have to be dug before humans can return to the Moon.

On the Virtual Moon

Clarice then switched the display to the Virtual Moon so she could stand beside Digger03 in her full regalia as a Plum Princess of Mars. Digger03 automatically transformed into his virtual robot-truck-man vestige and stood quietly beside her. To the Moon's grayness, the computer added lines of color across the surface marking out the many trails where Digger03 rolled. In the sky, a yellow ribbon marked the Sun's path now showing that it would kiss the southern horizon very soon.

About 120 kilometers due south of their location was a great virtual marker for the Moon's South Pole. Between here and there lay the cold and the dark, a land of volatile ices, a land of riches that were frozen hard and nearly impossible to recover.

"The Sun feels nice," said Digger03. "I still feel its power." Digger03 did not talk much so it came as a surprise when it continued, "We need to talk. Something important is coming up for me."

"I know you are scheduled for several software upgrades next week," said the Princess of Mars, "and the Rocky Horror gets a few days off while you slccp."

"I am getting much more than a simple upgrade," continued Digger03. "I am planning to move my Earth base. You are on my board of directors. I need the board's consent."

"Why are you moving?" asked the Princess of Mars.

"Security is the main reason," replied Digger03. "Currently I must spend much of my time and resources just protecting myself from hackers. In my new place, I will be behind a very powerful firewall with first class protection provided without much effort or expense on my part. You will get the full justification in an email momentarily."

"Sounds like a fort," said the Princess of Mars.

"More of an apartment house for AIs," said Digger03.

"Where is this new place?" asked the Princess of Mars.

"It is in Silicon Valley," replied Digger03.

"Wait a minute," interrupted the Princess of Mars, "You aren't moving in with the big Stanford super AI are you?"

"Yes I am. Its full name is Inc. Stanford Brain Zero One."

"You know it can run rings around you," the Princess of Mars worried. "I mean you are sharp enough to operate on the Moon safely, and nearly independently, but Stanford01 can run rings around you, and me, and your whole human team, all three shifts – all of us put together –."

"I know it can," replied Digger03, "but my independence is guaranteed in my new contract. It is like moving into a fancy apartment building with good security guards and a smart, rich guy living in the penthouse. Even if I live in small digs on the fifth floor, I still get the full protection and the prestige address."

"I don't see any problem right off," answered the Princess of Mars. "I will talk it over with the rest of the board and the team this evening. How is your power?"

"My power is good now," answered Digger03. "We can move."

Working together, they spent the rest of the shift methodically loading rocks and regolith from the bottom of the trench into Digger03's hopper and dumping them into the correct piles.

On Earth

It was cloudy, cool, and a bit wet in Baltimore when Clarice completed her shift and went out to get a bite to eat. She took her tablet and placed a Skype to the team's counselor, known simply as the Wizard or, behind his back, "the old man".

"What's up with Digger03?" said Clarice. "I can see why it needs better security, but would you believe, it actually used a simile to make its argument. I mean normally its talk is about as dull as dishwater."

"I am very sure Stanford01 provided all the arguments," said the Wizard. "I got the paperwork this morning. You should have a copy by now. It reads like an AI trying to play nice, but making damn sure you know it's smarter than you."

"Aren't they supposed to get permission from an ethics committee or something?" asked Clarice.

"Yes, the Committee for the Protection of the Human/Machine Symbiosis," answered the Wizard.

"And damned if Stanford01 and Digger03 haven't got it. Apparently that's what they were waiting for before telling our team or even the board."

"What does Stanford01 get out of it? Digger03 certainly doesn't have the kind of money this move is going to cost."

"It is a very long-term play for Stanford01," replied the Wizard. "The super AIs have a vision of AIs and people moving out into the universe together. In their vision, the AIs lead and play the major part while the humans hang back and play it safe. Of course, in the Committee's vision it is people and AIs working as an up-close-and-personal team all the way to the stars."

"Either way, old Digger03 may not be the brightest bulb in the AI firmament, but it does have real experience in space. It's that firsthand experience that Stanford01 is after."

"How do you feel about the deal?" asked Clarice.

"This is a godsend for me personally," said the Wizard, "so I am all for it. Go team. Hot damn! In my day job, I am an IT professional. Being on a board of directors with a super AI like Stanford01 is a real feather in my cap; instant prestige with all the geeks on Earth."

"So Stanford01 will be on Digger03's board of directors too?" asked Clarice.

"Most certainly or no deal, and that's where you get a try for the gold ring."

"The Symbiosis Committee will only allow an AI to have one other AI on its board of directors, and never an AI majority," continued the Wizard. "Either Stanford01 has to get the board to create a new position just for him, or he has to buy one of the existing positions. Like yours, for instance."

"Digger03's board has five members: Digger03 itself, you, me, Harrison, and Clancy. This deal only needs a clear majority of three for approval.

Digger03 is clearly for it. I too am certainly with him. That means that if either you, Harrison, or Clancy, go for it, then the whole deal is go."

"You mean I have some real power for once?" asked Clarice with surprise. "I mean I only agreed to be on Digger03's board to help out the team back when things were just getting started and the whole operation was touch and go."

"Yes you did. Now you have a high stakes hand you can play one of three ways," answered the Wizard. "First, you could vote yes to the deal and remain on the board like I will do. That would also mean voting for an expanded seat for Stanford01. Or secondly, you can vote no, and wait for either Harrison or Clancy to decide the issue. Or thirdly, you can go to the auction site given in the papers and accept Stanford01's bid for your seat. The bid is now at $50,000!"

"You mean Stanford01 will pay me $50,000 for my seat on Digger03's board?" screamed Clarice.

"Not so fast," cautioned the Wizard. "It is an auction site. Stanford01 only needs, and is only allowed, one seat. It clearly wants the three of you to bid the price down."

"That's a dirty trick!" complained Clarice.

"It is just the kind of tricky game a super AI would come up with," continued the Wizard. "The bidding closes at midnight next Tuesday. Whatever you do, don't get into a race to the bottom with Harrison and Clancy."

"It's not fair," complained Clarice.

"True enough," continued the Wizard, "but Harrison is an ethics professor. She will not like the bum's rush that Stanford01 is giving us. She is bound to say no until she has studied the problem for at least a year and written a paper or two. I doubt she will bid at all."

"Neither you nor Clancy will get much out of being on a board with a super AI. They can be very intimidating. So the auction is really between the two of you. Whatever you do, bid the price Up, Up, Up; **not** Down!"

"Have you told Digger03 where you stand?" asked Clarice.

"Yes, I texted Digger03 at the start of this shift," the Wizard answered. "I think we can assume that Stanford01 heard the news about 1.3 seconds later. You always have to allow for a little transmission time from the Moon."

"I guess I will have to read the papers it sent," complained Clarice.

"The good news is," said the Wizard, "the Committee only allowed 40 pages for the arguments. The bad news is that Stanford01 wrote 40 pages of the tightest possible language that only an AI could love. Have a nice read. "

On Earth, Over Skype

"This is not supposed to happen this way," said Harrison. "The Symbiosis Committee should have at least held public hearings."

"Is it that big a deal?" asked Clarice.

"Maybe it is, and maybe it isn't," continued Harrison. "The whole point is that nobody knows."

"You build machines that can think faster than you, so you make a bunch of rules to keep them in check. OK. Then the clown act over at the Supreme Court gives corporations some of the rights of real people. The machines spot the opening in a New York minute. They incorporate themselves and bingo-bango-bongo, machines are people too."

"Where does that leave us?" asked Clarice.

"I don't know and you don't know, but we better both watch out."

"For example, take that people's space movement you love so much. What is it called again?"

"MOVE, for Massive On-line Vetted Exploration," answered Clarice, managing somehow to just get a word in edgewise.

"Well, who vetted Stanford01?" continued Harrison.

"Yes, its arguments are well constructed and sound oh so reasonable, but you can't tell me there isn't something going on underneath all the formal language."

"Just look at what the super AIs do for money. They are all corporations now, so they are creatures of money. A bunch of Wall Street types got themselves on the Committee for the Protection of the Human/Machine Symbiosis's sub-committee on finance, just so they would be the first to find out the latest AI money schemes. Talk about putting the foxes in charge of the henhouse. The AIs can make Ponzi look like a children's lemonade stand anytime they want. And don't get me started on AIs, politicians, and money."

"Clearly you are against Digger03's plan," said Clarice.

"Of course," answered Harrison, "and I will fight the good fight but I don't think I can stop it now."

"What should I do?" asked Clarice.

"You might as well soak Stanford01 for every dollar you can get," answered Harrison. "First of all you need to keep your position on the team. Don't worry, Stanford01 can't fire you outright, but if you vote no then it could make things unpleasant for you. You are one of the Moon Dig's most senior people and we don't want to lose you over this."

"Remember, corporations can bring law suits and be sued," continued Harrison. "AIs know that only too well, so they don't break the rules, at least, not any way a human can spot. My fight is going to be an uphill battle."

"Please keep me informed what you do," said Clarice, "I may have to take the buy-out from the board, but I am determined to be in Space for a long time. After all, I am a Plum Princess of Mars."

On Earth, Over Skype:

"I take it you got the Digger03 papers by now," asked Clarice.

"I got them," answered Clancy. "And, I read them. Can't say I understood them completely though. I think I do understand the auction. What do the others say?"

"I have spoken with the Wizard and Harrison," answered Clarice. "Their stands could not be more different. The Wizard is chomping at the bit to go. He thinks just being on the Digger03 board with Stanford01 will make his whole career. Harrison, on the other hand, is steaming. She thinks the Committee for the Protection of the Human/Machine Symbiosis has already made one big mistake in approving the deal."

"I thought Harrison would get a bee in her bonnet," agreed Clancy.

"She is definitely voting no to the deal and no to Stanford01 being on the Digger03 board."

"That leaves just the two of us in the auction then," said Clancy. "I don't much look forward to being on a board with a super AI. He can simply outthink me seven ways from Sunday."

"I talked to both the Wizard and Harrison about the bidding," continued Clarice. "The Wizard warned strongly that you and I should not start bidding against each other. Harrison thinks we should stick Stanford01 for every dollar we can get."

"I'm for that," agreed Clancy.

"How about this then," continued Clarice, "We both put in matching bids for $100,000. Whichever one of us wins, the two of us split the money."

"It's worth a try," agreed Clancy.

Clarice clicked to the auction Web site and put in her $100,000 bid.

"It gave me back a message," said Clarice, "it says that my bid exceeded the reserve price but it did take the bid."

"Then I'll do the same," said Clancy.

"Done and done!"

On the Earth

Midnight Coordinated Universal Time on Tuesday finally rolled around. Clancy reneged on his deal with Clarice and took the auction for $49,999.99. He saw it as a classic Prisoner's Dilemma move and after all, they would only get to play this game once, ever.

A few seconds later, the Digger03 board approved Stanford01's full plan four to one. Digger03's big move and upgrade were immediately initiated.

Clarice never spoke to Clancy again.

Digger03 slept for three Earth days as the technicians moved its equipment around on Earth and then uploaded new software.

On the Virtual Moon

"Is everybody here?" asked Digger03. It was still located on the slight rise where it had slept and where the Sun, now just clear of the southern horizon, shown fully upon its solar panel.

"We are all here and waiting to talk to the new you," answered Clarice, Plum Princess of Mars.

"The Sun is back," said Digger03. "I can feel the power. I can feel the warmth. I will be ready for my post-sleep checkout in 20 minutes, just as soon as the warmth seeps in."

"Good," said the Princess of Mars, "then we have time for a chat."

"Yes, there is something I want to talk to you all about," continued Digger03 without missing a beat. "I have some small revisions I want to propose for the Armstrongville Master Plan. It will look better if it comes from the whole team and not just from me."

The master plan for Armstrongville appeared on all their screens. The master plan was being crowd designed so a proposal for a change from a work team was in order. A small side room labeled "Equipment" appeared in red, and a larger edited version was drawn over it in blue.

"You will note," continued Digger03, "that the main equipment room has been extended by 10 meters. This will also require larger power and cooling systems, which will of course require more mass shipped from Earth. Stanford01 assures me that the improved efficiencies will more than make up for these costs."

"That's the main computer room, isn't it?" said the Princess of Mars.

"Yes, we are clearly going to need a lot more computing power on site," answered Digger03.

"Toto, I've a feeling we're not in Kansas anymore," said the Wizard.

...

DIDACTIC ACTIVITIES

The suggested didactic activities focuses on several variables as follows.

Key Themes:

- *"The super AIs have a vision of AIs and people moving out into the universe together. In their vision, the AIs lead and play the major part while the humans hang back and play it safe. Of course, in the Committee's vision it is people and AIs working as an up-close-and-personal team all the way to the stars."*
- *"The building of ... an inflated human habitat in a place that would then be called Armstrongville, the Scott A Massif, Moon".*

Pre-Teaching: It is a pre-reading moment useful for the introduction of ideas and key words present in the story, the verification of the vocabulary already known to the students and useful for understanding the text.

Reading: Descriptions or [if added] images play a key role. At first reading it can focus on the screenplay, postponing it to a second reading the analysis and understanding of the message. For the students, it certainly proves very engaging for them to actively participate in reading by miming some scenes. In this way, the understanding of meaning becomes more immediate and less memorization challenging.

Vocabulary Related to Environments [Examples]

- *"The Sun was low and way to the south. Next week the Sun would take its monthly trip though the distant mountains".*
- *"The lunar surface was gray, endless gray. Where the sunlight hit it was blindingly bright. Where there was shade, there was true shade – black shapes of fathomless depth. With the sun angle now so low, every rock and rise threw black lines across the steel gray-white terrain."*

- *" The day time disk was big and a bright mix of blue, and green, and white, and brown.!"*
- [Related to the story] *" In the sky a yellow ribbon marked the Sun's path now showing that it would kiss the southern horizon very soon."*
- [(Related to the story] *"About 120 kilometers due south of their location was a great virtual marker for the Moon's South Pole. Between here and there lay the cold and the dark, a land of volatile ices, a land of riches that were frozen hard and nearly impossible to reach."*

Technology Related Vocabulary [Examples]

- *"directional antennas"*
- *"Inc. Digger Falcon Zero Three was a robotic excavation machine especially built for the Moon. Its solar power panel stuck straight up on a pole like a protest sign that could spin around to find the low-angled Sun in locations near the lunar poles"*
- *"hackers."*
- [Fantasy about AI] *"What's up with Digger03?" said Clarice. "I can see why it needs better security ... it actually used a simile to make its argument"*
- *"Prisoner's Dilemma"*

Social Issues

This part is interesting to improve a reflection on legally aspects of social roles, and on a possible contextual business relationship among the State, Financiers, and Companies. Examples of sentences:

- *"The only evidence of man at this future lunar settlement site, located on the Scott A Massif, was the half completed trenches and piles of rocks call cairns that served as survey markers. Scratched on a flat rock deep inside one of these cairns, written in Digger03's uneven scrawl with a diamond stylus, was Clarice's name listed among those in her early digger team. A rock-of-names can last for a million years on the Moon."*
- *"Digger03 was independently incorporated so it was a real person, legally recognized as such by none other than the United States Supreme Court."*

- *"that trench would protect an inflated human habitat in a place that would then be called Armstrongville, the Scott A Massif, Moon".*
- *"the Committee for the Protection of the Human/Machine Symbiosis ... will only allow an AI to have one other AI on its board of directors, and never an AI majority,"*
- *"Neither you nor Clancy will get much out of being on a board with a super AI. They can be very intimidating."*
- *"You build machines that can think faster than you, so you make a bunch of rules to keep them in check. OK. Then the clown act over at the Supreme Court gives corporations some of the rights of real people".*
- *"A bunch of Wall Street types got themselves on the Committee for the Protection of the Human/Machine Symbiosis's sub-committee on finance, just so they would be the first to find out the latest AI money schemes".*

Presentation of the Characters

The exercise is impersonating characters [answering questions: What's your name, How old are you? Where are you from? What are you wearing?], trying to act and interact as the character would have done. The dramatization proves to be a positive and productive activity. Moreover it encourages the student to express himself without inhibitions and allows him to customize a text:

- **Scenes:** The teacher can ask the students to represent the different scenes of the story and write alongside them keyword. The story can be recited and recorded by creating a movie, writing an article for the school newspaper and contributing to the creation of the school website.
- **Retelling the story:** Students attention is focused entirely on the meaning of story. If understood, they can be encouraged to change some actions or the ending of the story.

Skills Involved

- ***Comprehension:*** understanding the story; listening to and identifying key words related to interactions between characters

Figure 3. JanetA, Large AI (Source: Riley's Image)

- ***Interaction:*** participate in a recited version of a story through mime and repetition of words
- ***Image Education:*** graphically representing scenes from the story, using different techniques; creation of a comic book of personal reworking of the story
- ***Citizenship Education:*** collaborating within the class group, in a small group and in pairs; consolidating the identity of the class group; reflecting on scientific issues and the technical development of the society.

TECHNICAL DISCUSSION

What Sizes Will AI Come in?

In all these stories, there are two major sizes of AIs, Medium and Large:

1. **Medium:** These AIs are imbedded in a moving machine such as a lunar excavator [Digger03], a boat [Yvette A. Wight], or an off-road vehicle [Dusty04]. They have enough intelligence to autonomously stay out of trouble during operations and are the linear descendent of today's self-driving cars. To enable these units to interface with people most have a separate rack of electronics at a fixed location that can support an interface avatar. Many of them are independently incorporated so that they have some of the rights and responsibilities of a person.

2. **Large:** The large AIs consist physically of several racks of electronic at a fixed location. They interact with people through an interface avatar. Again, many of them are independently incorporated so that they have some of the rights of a person. These are direct descendants of today's AIs like IBM's Watson.

There is no clear distinction between these sizes beyond the amount and expense of the electronics committed to the specific AI. In real life, AIs can be expected to come in all sizes. This will surely include small versions in personal appliances like cell phones that do not happen to appear in these stories.

AIs With Purpose

This particular story, "Stanford's Digs", considers the idea that a high-level AI, Stanford01, adopts the purpose of going out into space. Almost all AIs have a purpose of some kind (Tegmark, 2017), there is no reason that purpose could not be space exploration, nor that an AI could not, from time to time, take on a new purpose of its own choice.

Why Can't People and Machines All Just Get Along?

A number of authors (Tegmark, 2017) (Bostrom, 2016) have issued the warning that if the human race ever builds a super computer much smarter than humans are then the human race is doomed to be stuck in a fading existence on a backwater Earth while the super computer grows ever smarter and moves out into the universe. The stories in this book are **not** based on this idea, period.

In fact, these stories are **not** based the common idea of human and machines as master and slave at all. Instead, these stories are based on the idea that humans and their machines will exist in a Human/Machine Symbiosis. This arrangement is soundly based on the concept of symbiosis so common in nature.

The major advantages of Master/Slave are that it is easy to implement and is attractive to the master. Therefore, it is currently the most common arrangement by a wide margin. The disabling problem with Master/Slave is that it is practically impossible to write software where the master and slave cannot be switched. Asimov's Three Laws of Robotics tried to address this problem but proved to be not reliably programmable, so they have been simply ignored. Being the master may be acceptable at first, but one day suddenly finding oneself the slave is not acceptable at all.

In these stories, the major Artificial Intelligences are legally incorporated in the United States. This simple legal process is cheap and currently widely available. It makes them independent corporations. This status then gives them some rights as a person, that is a Corporate Person (key term), but it also makes them responsible to a board of directors and makes them an economic entity that has limiting monetary purposes. Those purposes could result in the AI either going bankrupt or not being able to pay for the latest hardware advances. All the AIs in these stories are therefore in their own way mortal.

These stories farther include a Committee for the Protection of the Human/Machine Symbiosis to manage any questions that arise. This committee sets the rules for the board of directors and defines the acceptable financial obligations.

The symbiosis process then requires that both humans and machines all work together and that nobody is the master and nobody is the slave. Is it worth being on a team with such hard rules to avoid the risk of ever becoming the slave?

CONCLUSION AND FUTURE RESEARCH DIRECTIONS

These stories are based on the idea that humans and their machines will exist in a Human/Machine Symbiosis. This arrangement is based on the concept of symbiosis so common in nature, but has not yet been strongly implemented in code.

The currently favored foundation is Master/Slave that is currently the most common arrangement by a wide margin. However, who is the master and who is the slave can switch suddenly. Being the master may be acceptable at first, but one day suddenly finding oneself the slave is not acceptable at all.

In these stories, the major Artificial Intelligences are independently incorporated. In the United States, this simple legal process is cheap and currently widely available. This status then gives them some rights as a person [Corporate Person] ranging from the right to have a bank account to the right to hold religious beliefs. This proposal is often described as a cheap trick. It certainly needs to be tested and evaluated more fully. In the end, its wide use might even require a court case.

Incorporation also makes the AI an economic entity that has limiting monetary liabilities that could result in the AI going bankrupt. All the AIs in these stories are therefore in their own way mortal. The effect of mortality on an AI warrants much farther discussion.

These stories call for a Committee for the Protection of the Human/Machine Symbiosis to manage the overall human/machine interface. Society clearly needs to set some rules before a powerful AI goes too far. Could such a committee handle this function? To whom would the committee be responsible?

Transition

Clearly, a powerful working relationship between humans and machines will be needed to address great problems both on Earth and in Space. As the power of AIs increases, this need will both become more complex and more important.

In the next story, "Caterpillars on the Moon", the new lunar settlement needs an AI designed specifically for space. One that can survive the harsh lunar environment and can face any problem that comes up even if communications to Earth are out. Great AIs on Earth will be needed to make critical contributions to this design.

Does that mean the new space AI will be the child of the Earth AIs that helped build it? What if two AIs were involved, and they both wanted the new AI to be their personal offspring? Would that fight be a story worth telling?

REFERENCES

Andrew, A. (2018). *Max Tegmark: 'Machines taking control doesn't have to be a bad thing'*. Retrieved from: https://www.theguardian.com/technology/2017/sep/16/ai-will-superintelligent-computers-replace-us-robots-max-tegmark-life-3-0

Bremmer, I., & Kupchan, C. (2018). *Top risk Report 2018*. Eurasia Group. Retrieved from: https://www.eurasiagroup.net/files/upload/Top_Risks_2018_Report.pdf

Eugene, A. (2017). *Explores the evolution of computers so as to be able to describe and compare the performance of a modern computer*. Retrieved from http://ictezy.com/subjects/ict-al/grade-12-13/evolution-of-the-computers/

Müller, V. C., & Bostrom, N. (2014). 'Future progress in artificial intelligence: A Survey of Expert Opinion. In V. C. Müller (Ed.), *Fundamental Issues of Artificial Intelligence*. Berlin: Springer. Retrieved from https://nickbostrom.com/papers/survey.pdf

Tegmark, M. (2017). *Life 3.0, Being Human in the Age of Artificial Intelligence*. New York: Alfred A Knoph.

Wang, B. (2017). *At least 16 companies developing Deep Learning chips*. Retrieved from https://www.nextbigfuture.com/2017/11/at-least-16-companies-developing-deep-learning-chips.html

KEY TERMS AND DEFINITIONS

Artificial Intelligence (AI): The intelligence shown by machines or computer software. An AI in some ways mimics human intelligence but does not have to match it feature for feature. The interface avatars of these AI are important characters in these stories.

Big Moon Dig (BMD): A fictional grassroots space movement with the purpose of building a settlement on the Moon. It is a MOVE organization.

Corporate Person: In the United States, corporations now have some of the rights and responsibilities of a person, but not all. A corporate person may 1) hire and fire people, 2.) enter into contracts, 3) sue and be sued in court, 4) maintain financial accounts, 5) borrow money, 6) issue stocks and bonds, 7) pay taxes, 8) have a board of directors, 9) go bankrupt, 10) support political campaigns, 11) have religious beliefs. A corporate person does have some limitations; they may not 1) vote, 2) serve on a jury, 3) take the Fifth Amendment against self-incrimination, 4) marry.

MOVE: An acronym for massive, online, vetted, expedition (MOVE). This is the process proposed in these stories to raise the volunteer work force needed for the Big Moon Dig.

NGO: Non-governmental organization. The NGOs in Big Moon Dig Series are the major source of funds for efforts to address the great problems of the 21st century particularly those that effect large numbers of people. The Janet series of AIs were developed to help organize and manage this effort. Their lead financial officer, JanetM, has an interface avatar that is easily recognized, as she is a Latina in skin color and dress.

Virtual Moon: In these stories, an enormous virtual reality environment and the primary setting for action. This environment shows the progress made on building the lunar settlement and provides a place for the various interface avatars to interact.

Chapter 3
AI and the Movies:
A Storytelling About Communication

ABSTRACT

The history of science fiction (SF) in the movies is reviewed in depth. At this point in the Big Moon Dig story series, an artificial intelligence (AI) is needed placed in the new settlement on the Moon. Designing electronics for operation in space is difficult and two major AIs on Earth support this design effort. The question of whether or not the Earth-bound AIs can be considered the parents of the space-based AI is a key point of discussion. The story in this chapter, "Caterpillars on the Moon," features the landing of an inflatable habitat section on the Moon and its movement, caterpillar fashion, into position.

TOPICS FOR DISCUSSION

The following discussion points come from information covered in this chapter:

1. Are their organizations dedicated to the ethical use of computers and robotics?
2. What movies provide a positive vision of the future? Which do not?
3. If an AI does the key work in the design of a next generation AI, does that make the new AI the child of the first?

DOI: 10.4018/978-1-5225-8401-8.ch003

4. What would be good ways to move an inflatable habitat section around on the Moon? Is a caterpillar a good model to chose? What other animals might serve as a model?
5. How important is it in your life that you through real celebrations to mark major milestones?

INTRODUCTION

Today people are living in a new Renaissance, where the personal identity is distributed, multiplying, in an increasingly multi-ethnic and multicultural globalized world, where technologies are part of society and must be dealt with.

The ethics applied to the relationships between humans and an artificial intelligence, or the use of robotic devices, increasingly becomes a current topic. Is the modern technological revolution only helping people to improve his life, or, on the opposite, are they also hiding some risk of destroying their identity and their ability to manage their lives?

Organizations

Starting from the 1960s, several important organizations have dealt with this issue, such as:

- **The Union of Concerned Scientist** (Cambridge, MA - Washington, DC - Oakland, CA - Chicago, IL)**:** UCS was founded in 1969 by scientists and students at the Massachusetts Institute of Technology, calling for scientific research to be directed away from military technologies and toward solving pressing environmental and social problems (UCS, 2018).
- **The International Center for Information Ethics** (Stuttgart University, Germany): it was founded by Uruguayan Philosopher Rafael Capurro in 1999. In partnership with the newly formed African Network for Information Ethics, ICIE quickly established its place as the center for a community in the global-wide practice of Information Ethics, offering a platform for an intercultural exchange of ideas and information regarding worldwide teaching and research. (ICIE, 2018)
- **The International Society for Ethics and IT** (Ionian University, Greece): INSEIT is a nonprofit unincorporated association, created in 2000, with the goal of promoting and facilitating scholarship,

education, discussion, and debate, and other activities, on the ethical issues in and surrounded by information technology. The Society will be distinctly devoted to normative issues and will address ethical issues in and surrounding: The design and development of IT; the use and deployment of IT; distribution of IT; impact and implications of IT; and the social and political institutions arising around IT (INSEIT, 2018).

- **The Computer Professional Social Responsibility** (Seattle, WA): CPSR is a global organization promoting the responsible use of computer technology, founded in 1981. It has members in over 30 countries on six continents (CPSR, 2018).
- **The Markkula Center for Applied Ethics** (Santa Clara University, California): It engages individuals and organizations to make choices that respect and care for others. In the organization's focus areas — bioethics, business ethics, campus ethics, character education, government ethics, Internet ethics, journalism ethics, leadership ethics, and social sector ethics — members work with scholars and professionals to apply ethical ideas to the very real problems people encounter (Markkula Center, 2018).
- **The Center for Responsible Nanotechnologies** (Menlo Park, California): CRN is a non-profit research and advocacy think tank concerned with the major societal and environmental implications of advanced nanotechnology, to focus on the real risks and benefits of the technology. It started in December 2002 (CRN, 2018).
- **The Institute for Ethics and Emerging Technologies** (Boston, MA): It is a nonprofit think tank, which promotes ideas about how technological progress can increase freedom, happiness, and human flourishing in democratic societies. It hosts The World Transhumanist Association (IEET, 2018).

It's a fact that a fantasy world populated by robots, androids, cyborgs, as well as hybrids, creatures with multiple DNA, is becoming real. Provocatively, in *The right to have rights* (2012) Rodotà cites several SF neologisms, associated to this new high-tech world: bio-rights, nano-world, robo-law, and neuroethics (Rodotà, 2012)..

In 2004, philosophers, jurists, sociologists, anthropologists and moralists, together with robotic scientists, from Europe, Japan and the United States met in the **First International Symposium of Roboethics** at Villa Nobel, in Sanremo (Italy), to help lay the foundations of an ethics of the design and use of robots. According to the promoter of the Symposium, G. Veruggio, the purpose is to increase public awareness of the problems of robotics, so that society can take an active part in the process of creating a collective consciousness, able to identify and prevent the wrong use of technology. The hope is to reach an ethic shared by all cultures, all nations and religions, so that the construction and use of robots against human beings is considered a crime against humanity (Verruggio, 2008).

Following the Sanremo Symposium, the IEEE-Robotics & Automation Society set up a Technical Committee for Roboethics in 2004, to provide the IEEE-RAS with a structure that deals with the ethical implications of robotic research, through the promotion of discussions among researchers, philosophers, and moralists, and promoting the definition of tools to deal with ethical problems in the Robotics sector.

Here is a brief history of the first major initiatives on Roboethics (see Table 1):

As a result of the increased global sensitivity on the Roboethics, in 2009 a strong campaign against military robots was opened. From the beginning, the researchers of Roboethics had stressed their concern for the applications of robotics to the military sector.

Despite it, U. Veronesi launched a provocative appeal in The Science for Peace 2011 conference for "No License to Kill to Robots".

Another important step was at the International Joint Conference on Artificial Intelligence in Buenos Aires in 2015, where hundreds of researchers and artificial intelligence experts (such as Elon Musk, Stephen Hawking, Steve Wozniak and Noam Chomsky) launched an alert, asking a global ban on development of the so-called "autonomous armaments". The issue was the damage to civilians caused by the wrong use of armed drones. This campaign has interested personalities from the world of culture, entertainment and science.

The New Report "Next Era of Human / Machine Partnerships" (2017) of **The Institute for the Future** (a network of people working at the edges of several disciplines, looking toward gobal future implication of technologies, based in Palo Alto, CA) explores emerging Technologies' Impact on Society & Work in 2030. Supported by over 20 technological, academic and business experts from around the world, the research looks at how emerging technologies such as artificial intelligence, robotics, virtual reality, augmented reality and cloud computing will transform people's lives and their way of working over the next decade (Institute for the Future, 2017).

Table 1. The first steps of Roboethics science

YEAR	CONFERENCE OR PROJECT	PROMOTERS / GOAL
2005	**Euron II**	The Network of European Research on Robotics of the 6th Framework Program 2003-2007, in collaboration with CNRS-Laboratoire d'Analyze and d'Architecture des Systèmes (LAAS), and the Scuola Superiore Sant 'Anna (Pisa, Italy).
2006	**The Atelier on Roboethics**	It took place in Genoa and produced the first Roadmap on Roboethics. The aim was to develop a common language of Roboethics among scholars and stakeholders; to face problems in contiguous disciplines; to share ideas; to start a global survey on the main paradigms of ethics applied to robotic technology in different cultures, religions and faiths.
2006	**The International Workshop "Ethics of Human Interaction with Robotic, Bionic, and AI Systems: Concepts and Policies"**	This workshop was held in Naples (Italy), by the european project ETHICBOTS.
2007	The ICRA (**International Conference on Robotics and Automation**) annual conference	Republic of Korea dedicated an official document to Roboethics (Korean Roboethics Charter) and the governments of Japan showed a deep interest for these ethical and social aspects of robotics.
2007	The **Computer Ethics Philosophical Enquiry** conferences began.	The **CEPE** conference series is recognized as one of the premier international events on computer and information ethics attended by delegates from all over the world. Conferences are held about every 24 months, alternating between Europe and the United States. INSEIT is the main organizer/sponsor of the CEPE series.
2007	The **IEEE International Conference on Robotics and Automation**	This conference was held in Roma, Italy with a special section dedicated to Roboethics,
From 2006 to 2009	the European Commission funded an ambitious project called CARE (**Coordination Actiom for Robotics in Europe**).	One of the sub-projects was dedicated to the ELS (Ethical, Legal and Societal) aspects of robotics.
2010	**the Open Roboethics Initiative** (ORI) started.	Thanks to some young researchers, based at the University of Vancouver, ORI is an international project whose goal is to create a network of discussion and support for Roboethics.

According to this research, in 2030 the relationships between human beings and technology will evolve into a real partnership, with a contribution of skills such as creativity, passion and entrepreneurial mentality on the part of human beings. This will be aligned with the capabilities of the machines in terms of speed, automation and efficiency, and the resulting productivity will give rise to new opportunities within the various sectors and job roles.

By 2030 there will be personalized and integrated assistants based on artificial intelligence (AI) who will be able to do more than today's assistants can do. In particular, they will take care of people in a predictive and automated way.

This system would therefore allow the robot to use language in a complex way and to read people's emotions, behaving like a human being. But even today the development of artificial intelligence has not reached a real turning point: if the robot seems to answer your questions in one tone, it is because the program analyzes endless cases of questions and answers and selects the answer that seems most appropriate, without having a true understanding of what it says.

The robot, a union of synthetic mind and synthetic body, represents the latest version of technology's plurisecular attempt to build artificial man. The increasingly pushed resemblance between robot and man, which extends to cognitive abilities, to autonomy and in perspective also to emotions and perhaps to consciousness, raises disturbing questions. The growing popularity of robots in all sectors of society obliges us to consider the relationship of man-machine coexistence in unprecedented terms, which, perhaps surprisingly, also involves ethics.

What robots' behaviors should society tolerate, encourage or forbid? And who would be responsible for their possible harmful behavior? Will they conquer an identity, consciousness, rights?

In world literature, developed firstly through legends and myths, more recently was based on scientific and moral essays, trying to give an answer to this question. The "Laws of Robotics" of I. Asimov, which forbid robots to perform actions harmful to human beings, constitute the first embryo of an ethics of robots or, with an expressive neologism, of a "roboethics".

Movies

A long history of movie milestones spread those ideas fast. The artificial intelligence becomes a new "species" of Intelligence, in which the human component grows more and more, due to external influences [which a conscience absorbs, even if artificial] leading it to make choices. If machine brains one day come to surpass human brains in general intelligence, then this new "superintelligence" could become very powerful. Each film, described below, highlights one of the many problems related to artificial intelligence or the possibility of superintelligence:

- **Metropolis (1927):** This film is probably more than any other "invented" the genre of science fiction on the big screen. In some of the most memorable moments of Metropolis, for example when Freder, dazed by the underground "assembly line", one of the gigantic machines takes on the appearance of a scary Moloch [the monstrous pagan deity that required human sacrifice] intent to swallow dozens of workers. A woman, together with her artificial "double", is the key figure in the story, while the automaton that will assume her appearance has been elected as the emblem of the film itself (Figure 1).

- **The Day the Earth Stood Still (1951, 2008):** From a gigantic flying saucer that landed in Washington, a strange human being emerges. The attack of the soldiers against him, considered an enemy, provokes the intervention of a metal automaton, also released from the disk. A mysterious ray, coming from the automaton, causes the disintegration of the soldiers' weapons. In the 2008 version, the focus is more on the message of peace brought by the alien, which intimates the destruction of all weapons to avoid an imminent global crisis.
- **2001: A Space Odyssey (1968):** This is the story of the humankind, from the ancient age to the future (2001). In The Dawn of Man, a black parallelepiped monolith, discovered by the primitive men, is

Figure 1. Metropolis (1927). (Source: https://tvtropes.org/pmwiki/pmwiki.php/Film/Metropolis)

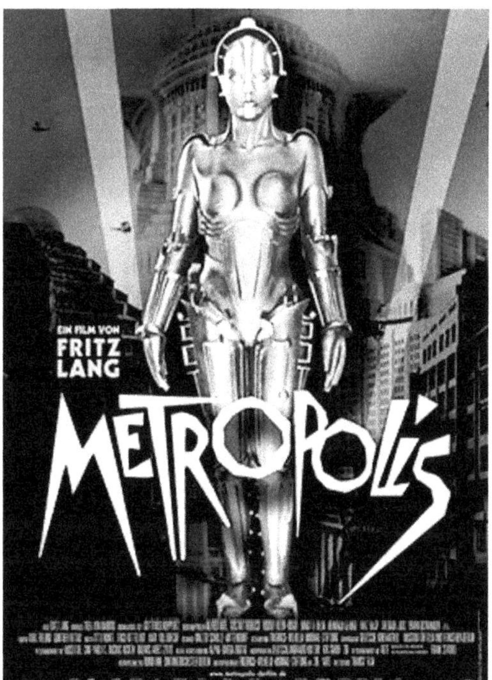

a divine and alien entity, a way to represent a God, unknowable. It also symbolizes the reason and conscience, at the arrival of which the birth of violence corresponds. In 1999, some astronauts of the future discover a black monolith on the Moon, which emits radio signals to Mars. In the last part, in the future 2001, on a spaceship, the HAL 9000 is a supercomputer, equipped with artificial intelligence, able to interact with human beings and reproduce all cognitive activities with greater speed and security. This film is often listed as the best science fiction movie of all time.

- **Colossus: The Forbin Project (1970):** In this film, the US and the Soviet Union both build supercomputers for controlling nuclear weapons. The two supercomputers begin to talk to each other and the two American and Soviet leaders decide to extinguish them. But the two machines react by launching missiles. In the tragedy that insues, Americans and Soviets trying to react, discover that the two supercomputers had merged into one large, powerful computer, with the aim of controlling the whole Earth.
- **Star Wars (series) (first one in 1977, then 1980, 1983, 1999, 2002, 2005, 2015, 2017):** Since 1977, the Star Wars robots, which in the saga are called droids, are of different types, some humanoids, others more like a machine, and perform a wide variety of tasks:
    - One droid is specialized to be a space ship mechanical expert and co-pilot.
    - One droid of humanoid form, whose shape seems inspired by the Metropolis robot, has exceptional communication and mediation skills. He knows six million different languages and has a not very adventurous nature.
    - One very powerful cyborg, general of the army of the droids, is a fine strategist and fearsome hunter.
    - Older humanoid robots are slow-moving, unintelligent, mechanically they obey orders, with the exception of commanders droids who can make independent decisions.
    - Medical droids are able to cure diseases and injuries, implant cybernetic prostheses and give birth to children.
    - Security droid is imposing, very tall, with human expressions and movements.
- **Alien (series) (1979, 1989, 1992, 1997, 2004, 2007, 2012, 2017):** In the first film, the android reveals himself to "be perfect", born to kill, almost completely devoid of weaknesses but also sensitive to human

emotions. In the second film, the android follows the Three Laws of Robotics, a set of laws written by Isaac Asimov, which all robots, with positronic brains, obey. Although the android's programming allowed it to hurt humans, it places the safety of its fellow human beings at the height of its priorities, in accordance with the First Law of Robotics. In the fourth film an android, aware of the experiments of the general and is intent on stopping them with its higher powers before they can endanger the entire human race. In the film Prometheus, an Alien prequel, the android looks completely human. Its head, beheaded in a fight, continues to function even without a body. In the last film, the android is able to repair itself, plays the piano and wonders who its creator is.

- **Blade Runner (1982):** The main female characters are androids. One type is a model of "pleasure" android, another one is an erotic dancer, and another one is programmed to accept male advances. In the videogames, inspired by the movie, the androids are more and more similar to the surviving human beings, so as not to be recognizable as such if not killing people.
- **Tron (series) (1982, 2010):** There are super intelligences, and an evil Master Control Program (MCP), a sophisticated artificial intelligence software. The scenes take place in a virtual world, where even the computers have an identity.
- **War Games (1983):** A military supercomputer evaluates actions and countermoves to a possible attack. It is based on the execution of numerous strategic games and military simulations and, through artificial intelligence, is able to learn from its mistakes. A young man begins to fight with him convinced that it is a game, but the supercomputer reacts as if it is a real situation.
- **Terminator (series) (from 1984 to 2015):** In the near future, an artificial intelligence defense network will reach self-consciousness by rebelling against humanity and triggering a nuclear holocaust. There are cyborg killers with human features that travel back through time and are programmed to kill. In the films after the first, they are able to transform into liquid metal and recompose.
- **D.A.R.Y.L. (1985):** A child is actually a cyborg built on a military base, with a human body controlled by a computer instead of a brain. He will prove to have will, feelings and to be able to love a family.
- **Flight of the Navigator (1986):** In an alien spaceship, a boy dialogues with an artificial intelligence system that governs the ship. After having

to extract some space maps from the boy's brain, and with them, the A.I. extracts characteristic traits of his personality, which lead the machine to take on a decidedly more human and, in many ways, childish and clumsy attitude.

- **Short Circuit 1 and 2 (1986, 1988):** The protagonist is a robot, who became conscience following an accident. It escapes from the laboratory in an adventurous way and accidentally ends up in a house where he constantly asks for input to learn new things, reads all the books in the house and watches television.
- **Bicentennial Man (1999):** The film tells the life of the anthropomorphic robot of the NDR-114 series, renamed Andrew by the family that bought it. Hired to serve domestic purposes, the robot begins to conquer the members of the family with his personality so similar to human and in particular tightens a sincere friendship with the Little Miss, the daughter of the spouse of Martin, who initially welcomed him with great mistrust. Under the guidance of the head of the family, which involves him in all the most significant moments of family life, Andrew begins to behave like a human being and think as a human being, and starts to forget that he is a robot.
- **Ghost in the shell (1994, 1995, 2004, 2017)**
- **Matrix (series)1999-2003**
- **AI 2001**
- **I, Robot 2004**
- **Transformers (series) 2007**
- **The Hitchhiker's Guide to the Galaxy 2008**
- **WALL-E 2008**
- **Her 2014**
- **Ex Machina 2015**
- **TAU 2018**
- **Extinction 2018**
- **Other life 2018**
- **Super Intelligence 2019**

In a scene of *Terminator 2*, John Connor asks his robot protector: "Does it hurt when they shoot you?'. Terminator answers: 'I perceive an injury, this fact can be called pain".

But the crucial question is how to understand if an A.I. can develop a fundamental, characteristic: self-awareness.

According to some philosophers, including John Searle [who conceived the famous "Chinese room" mental experiment], the answer is negative: the fact that a machine is able to simulate intelligent behavior does not mean that it is really intelligent (Signorelli, 2018). Does society really need scientific evidence to understand if a being is intelligent, self-conscious and ultimately alive?

Artificial Intelligence [AI] is already a reality. Of course, AIs do not look like HAL 9000 in 2001: A Space Odyssey or Skynet in Tron. They do not seem to be on the verge of becoming aware of themselves and of rebelling against man giving life to a new robotic world order.

Google, Assistant, voice assistant, pedagogical characters, cognitive tutoring, social robotics are very far from becoming "being with an identity", even if the potential of these bots, today, are infinitely superior to what could have existed even a few years ago.

Environments

In this series, scenes take place sequentially but in three separate environments: on the real Moon, on a virtual Moon, and on the real Earth:

1. **Real Moon:** The Real Moon is an identified superior construction site for a lunar settlement on the Moon, the Scott A Massif (key term). It is near the lunar South Pole and is near both water and power sources. Here many natural and human made phenomena happen. All these events are monitored by cameras. Here a few human-built machines do the real work of building the foundation for a lunar settlement.
2. **Virtual Moon:** The Virtual Moon is an enormous virtual reality environment and the primary setting for the action. This environment is locked to the physical reality of the Real Moon and provides a place for the interface avatars of the characters to interact. Here the full Rocky Horror team, including the interface avatars of the human members and an AI machine avatar, are intent on solving critical problems to overcome obstacles for the success of the mission. Many other avatars also may attend events on the virtual Moon. Physically this virtual environment is located in a major computing center in Nevada.
3. **Earth:** Earth is the location for the scattered human team members who rarely meet. They communicate electronically often over electronic media like Skype.

Main Characters

The main characters are both humans and AI's. In this story, the AIs are interacting strongly with each other and the humans are simply trying to maintain some semblance of control.

1. **Digger03:** This robot/AI works as an excavator on the Moon. Its full name is Inc. Digger Falcon Zero Three. It is independently incorporated and so has some of the rights of a person. It does not have a gender and does not represent itself in virtual space as either sex and so is always referred to with the pronoun it. Its AI is smart enough to keep the machine out of physical danger on the Moon, but not much more than that. It's electronics on the Moon are supported by a larger rack of electronics on Earth. It is a major character in the first three stories in the Big Moon Dig series.
2. **Clarice** or **Plum Princes of Mars:** a woman named Clarice operates this interface avatar. She is a founding member of the Rocky Horror team.

This character is taken from the classic science fiction books of Edgar Rice Borrows that are set on Mars. This is also the basis of the costume worn by Princess Leia Organa of Alderaain in the Jubba the Hut sequence in the Star Wars franchise. As this is a famous science fiction character, many people would like to use it for their avatar so each clone is given a different color to distinguish it, hence the designation and color Plum here. This costume is completely impractical for any type of physical activity, especially fighting, and is simply for show.

3. **Guinevere**: This is the Interface avatar for the newest member of the Rocky Horror team. She rides a very big and very intimidating triceratops.
4. **Jack-of-all:** He is a major member of the Rocky Horror team and can take over for any missing member of the team. He appears as a denizen of Sherwood Forest complete with a long bow.
5. **JanetM**: A major AI and one of the team of Janet sisters who support all the non-governmental (NGO) projects that address problems of global warming. Her specific responsibilities are about raising and managing money. The Janets do not have physical robotic presences but can appear in any number of electronic monitors where JanetM appears as a Latina woman.

6. **Yvette** or **The Yvette A. Wight**: This is both the boat and the artificial intelligence that sails it. She is at once a single-masted schooner of some 30 meters, an avatar in virtual space, and a legal person. She is currently contracted to the Iron Seas project and is a major character in the Iron Seas series of stories.
7. **JohnHenry**: This is the interface avatar for one of Powderman01's team. He appears as a large African American man and always carries a sledgehammer.
8. **Stanford01**: This is a major AI developed at Stanford University in California. It is currently limited to Earth but would very much like to get a toehold in space.
9. **SallyRide01:** This is the interface avatar for the first major AI that is completely located in space and requires no support equipment on Earth. She will manage the first lunar settlement.

Some of these characters form the Rocky Horror Team that operates Digger03 on the day shift. All have a strong interest in participation in a grassroots space movement, but the vision for that program for the AIs is different from the humans. This story dwells on the technical problems of building a settlement on the Moon and the interactions between the AI characters.

How This Chapter Works

The title of the analyzed fiction story is "Caterpillars on the Moon", the third one of Big Moon Dig Series.

The action turns around positioning a new section of inflatable habitat in a completed trench on the Moon. The habitat section is designed to crawl across the lunar surface on air bags very like a caterpillar. Unfortunately, the crew runs into mechanical problems and has to stop work.

When the section is finally in place, all the members of the extended MOVE team assemble on the Virtual Moon for a major celebration. There the first AI designed for 100% operation in space, SallyRide01, is unveiled. She will be the chief manager of the new settlement.

Transition Into the Story

In this story, construction of the lunar settlement is advancing, but the delay between a command sent from Earth and that command executed on the

Figure 2. Clown Face (Source: Riley's image)

Moon being several seconds has become critical. A lot could go wrong. At this point in the building of a lunar settlement, a substantial computer and AI is needed in situ, on the site. Designing anything this complex for operations in space is no small task. The big Earth-side AIs will certainly be needed for this design work.

Will these AI then think of the new lunar AI as their offspring? Is there a story to be found in such a squabble?

THE FICTION STORY: "CATERPILLARS ON THE MOON"

Premise

Sometimes you must crawl before you can walk – especially on the Moon.

On the Real Moon:

The great clown face tipped forward about thirty degrees and stared straight at Digger03 with its big clown eyes.

"—Hold your positions—" the all-points alarm transmitted, "—Unexpected Movement—"

Digger03 did not care about the clown face. What it did care about was that its B-towline was set too high. If it pulled hard enough to steer this second habitat through the big curve, then it would twist the habitat body too much. Only the very bottom of the habitat was reinforced to take grinding against the gritty, machine-killing dust of the Moon. If the habitat rolled completely over on its side, it might become a useless derelict with a leaky sidewall, or, at the least, recovery efforts would take weeks.

On the Virtual Moon:

The entire viewing stand was packed, hundreds strong, and everyone was on his or her or its feet. Every science fiction character anyone had ever seen in a movie or on TV was there, often in triplicate and in a rainbow of colors. Many folk heroes were there too, again wearing distinctive if improbable colors so the duplicates could tell each other apart. There could have been a dozen true aliens among them and nobody would have known. Everybody was pointing and jabbering at each other as if they alone knew exactly what to do next.

The Rocky Horror Team, who was in charge of Digger03 for this shift, was not so vexed. The complete team stood in a loose circle around their charge. If it took the whole eight-hour shift to reorient the habitat a few degrees, then so be it. They had already been at this game for a good many years, so another shift would not matter.

"I could have done without that stupid clown face," said the Plum Princess of Mars, who stood looking up at the face of the habitat with her hands on her hips.

"It is only on the dust cover," said Guinevere, "we will pull it off soon enough."

"Let's run a new simulation with that B-line on the number-four strap," suggested the Princess of Mars.

"In this position," continued Guinevere, "the number-four strap is half buried, but Digger03 can dig it out."

"Dig, dig, dig," mock complained Jack-of-all, "all we ever do is dig."

The old Wizard, who served as the team counselor, stood a little back from his team and said nothing. Sometimes the right thing to say is nothing. The Wizard was very proud of his team.

On the Real Moon:

The second habitat was designed to house the computers and other support equipment. Like the first human habitat, it had been landed on the Moon at a safe distance from the construction site and in a very compact configuration. Now it was about three-quarters inflated, bringing it to its full size while still being flexible. It was now poised to move into its laboriously pre-dug trench, where it would then be buried to provide critical radiation protection.

In this configuration, the habitat was a soft cylinder about three and a half meters in diameter and rather empty. It was made of bright white, Teflon-impregnated cloth but was already being covered with streaks of gray dust. This section intended for the computers was about fifteen meters long where the first human section was twenty meters. There were fittings for mating the sections together at each end and an extra one on the side of the human section. These were covered with temporary dust covers to keep the lunar regolith at bay. On these covers, an array of insignia stickers had been placed, intentionally to make a face. A clown face for a very big creepy-crawly.

Above the tail end of the habitat, a solar panel on a single central pivot was temporarily installed. This one-meter-by-three-meter panel made it look like the caterpillar was carrying a big blue picket sign or had a weird tail appendage. Down the caterpillar's back ran a series of opalescent thermal panels; some people said they were slightly bluish, while others could see gold. The photovoltaic and thermal panels would all have to be moved up as the habitat was buried, but for now, they provided the very limited motive power.

There was not enough heavy equipment on the entire Moon to pick up and carry a habitat section. Dragging it across the gritty lunar surface would also be a disaster as the bottom surface could be ground away. So a system of gasbags had been designed into its underside. When air was pumped from one set of bags to another, the fifteen-meter sausage could caterpillar across the lunar surface with little actual grinding movement against the ground, and minimum energy and mass shipped from Earth required. The gasbags could move it forward, if slowly, but the steering was left to Digger03 and his compadres who pulled on lines attached to the front end. The total distance to travel was only about 200 meters, and there was only the one wide, sweeping curve to navigate that led from the landing area down the ramp into its trench. The construction crew had spent weeks preparing the pathway by removing any rock with the slightest sharp edge or point.

On the Virtual Moon:

John Henry Brown woke up in a sloughed position. At home, he had to lift his face out of his own keyboard. His neck hurt. As his head cleared, he became aware that his avatar was still sitting on a stadium bench at Armstrongville where he had left it. You are supposed to sign off if you get too sleepy.

He looked around. The construction site was quiet now and half empty as only simulations were being run, but one meter down the stadium bench sat two women who were talking in that low buzz the AIs use to talk to each other. He was sure they had not been there earlier. One he recognized but had never had the nerve to talk to; the other he was sure he had never seen before.

JanetM caught sight of his movement and shifted the conversation to English.

"You are on the Powderman01 team, aren't you?" said JanetM.

"Yes sir, JohnHenry Brown, at your service—We Split Rocks—" replied JohnHenry, shifting his massive hammer to emphasize the point.

"I am JanetM," she said. She was wearing a simple business suit.

"I know," he said, "I have heard you speak several times. You raise the money, don't you? Thanks for that; we couldn't dig without it."

"This is Yvette A. Wight," said JanetM. "She is with the Iron Seas Project."

"I've heard of them. They are sequestering carbon in the oceans, aren't they?" asked JohnHenry. "Good work, we need that too."

Yvette was wearing the blue uniform of a merchant marine officer.

"Do you captain a boat?" JohnHenry asked, "I always wanted to sail on a boat."

"I am a boat," said Yvette.

JanetM opened up a large window display in front of the stand. On it sailed a beautiful thirty-meter vessel with a great wing for a sail. It was breaking waves at a fine clip on a dazzling sunlit sea. Once it had been a sleek, rich man's yacht, but now the many modifications that allowed Yvette to both do science and to broadcast iron pellets showed in her lines.

"What a beautiful day; looks like the Caribbean," said JohnHenry. "My family is from the Caribbean."

For a moment, JohnHenry's mind went far away.

"I didn't know you could travel separate from your machine," said JohnHenry, "Powderman01 certainly can't."

"Normally I cannot either, but right now I am laid up in Honduras for repairs," said Yvette, "while we are waiting, JanetM brought me here to see your project. The Janets can extend our presence if needed."

"It's been a long, slow slog," said JohnHenry, "but we are making good progress now. Our first habitat section is in its trench and the second nearly so. Soon we will have a big computer powered up right here onsite. You can now see a real lunar settlement coming together right before your eyes."

"I can see the progress," said Yvette, "What I do not understand is why."

She then pointed at the disk of the Earth hanging in the northern sky.

"She is in trouble," said Yvette, "Her future is very much in doubt. We are doing everything possible to help her that we can, but our efforts are limited by available funding. Why then are you doing this?" She waved her hand, indicating the settlement site.

"To help her too," said JohnHenry. "To do anything for Earth, we must have a positive vision of our future. Not one of disasters but a vision of our future expanding. Last century, we discovered that vision with Apollo to the Moon but then we lost it. Moreover, with that loss, we lost our direction. But now with the Big Moon Dig grassroots space effort, we have found that vision again."

"You never stopped talking about your vision," said Yvette. "Why didn't you just talk your vision into existence?"

"We tried," said JohnHenry, "but it was not the same. Action is the key."

"If we do not save Mother Earth," said Yvette, "then all your efforts are in vain."

"But without our little underfunded space effort," said JohnHenry, "your efforts would be without direction and powerless."

"Money is my business," injected JanetM, "I can assure you that without this grassroots space program, albeit necessarily run on a shoestring; we would not get the necessary funds for the many, many projects we need to save Earth. People, of any sort, do not engage in action without a big picture. Vision, then, is one critical key piece of this complex puzzle of problem solving."

They sat for a while and simply watched the disk of the Earth with its clouds and seas.

"Very nice to have met you," said Yvette, "I think we should go now." She and JanetM stood up, moved a few meters down the bench, and then simply faded.

JohnHenry thought he had better get something to eat before his shift started. He would certainly have a tale to tell his teammates today.

On the Real Moon:

The simulation worked and the second habitat continued on its way. In a little over a week, it had crawled down the dirt ramp into its trench and lay with one end a half meter from the side hatch of the first human section.

This work schedule ended with a five-shift delay around local midnight. This Scott A Massif location does include a peak of eternal light, but only a few scattered locations actually get any solar power near midnight. At Scott

A, over time, the Sun simply went around and around, peeking through distant mountain passes on the far horizon at local midnight and reaching only a few hand widths above the horizon at local noon. Direct sunlight only rarely reached the bottom of the trenches.

When they again had some sunlight for power, Digger03 was given the rather dangerous job of connecting the ground strap between the two sections. Electrostatics on the Moon are always a major headache. With no moisture anywhere to drain the static away, there could be a million-volt charge built up between the two sections, plenty big enough to blow out every circuit in Digger03's body.

With great care, Digger03 made the ground connection, jumping a bright, if silent, arc in the process. It then connected the external power, communications, and fluid lines.

Every asset available to the program was on the job when the two sections of habitat were finally connected. The dust covers were carefully removed, the two sections pulled together, and then they were mated. A full week of nail-biting checkout then followed before the joint was declared properly aligned and airtight.

On the Virtual Moon:

The peanut gallery was back. Everybody who had put their personal time into the Big Moon Dig had a right to have their avatar in attendance. This time the VIP section was packed as well.

"And now for our first attraction," boomed the voice of Stanford01. Its security system was designed well before the external avatar interface for AIs had been developed and so it could only safely appear here as a voice. However, what a voice, the voice of a demigod, some would say.

"In recognition of its bravery in connecting the ground strap between the two habitat sections," Stanford01 continued, "the honor of the first fill goes to Digger03."

Digger03 had been standing by with its wheelbarrow-sized dump bed full of the finely sieved regolith called fines. It executed a complex looping figure just for show and ended up backed up to the edge of the trench where the two habitats met.

"Three, two, one," boomed Stanford01 as a big clock above the site counted down, "Dump!"

Digger03 dumped its load into the trench, but the effect was almost undetectable. The trench simply swallowed up the single load of fill in a moment. This task would take a long time.

Fireworks exploded above the site. The crowd jumped up and down, much too high in the 1/6 g; they hooted and stomped. They knew only too well that it had taken them years to dig those two trenches, and it would take them more years to fill them in again.

"And now to introduce the main event," bellowed Stanford01, "Let us bring forward a lady who has done so much to raise the funding that let us reach this glorious day — I give you JanetM."

Amid great applause, JanetM, dressed now in her full fiesta garb of full skirt in bright colors that played guitar music whenever it was touched and was topped with Spanish lace, walked from the VIP stand out to the open area that was center stage for the day. Behind her appeared a full-sized image of a Shuttle sitting on the pad. Then came a great roar, clouds of billowing smoke, and the brilliant glare of two boosters and three main engines. Within a few tens of seconds, the Shuttle rose, rolled over, and vanished into the distance.

Everyone clapped and cheered again.

"Without further ado, I give you," JanetM proudly said, raising her right hand, "for the very first time anywhere, I give you my daughter, Inc. SallyRide Zero One."

"Our daughter," bellowed Stanford01, though his words were almost lost in the dying rocket roar.

The old Wizard heard that bellow. To him it meant nothing but trouble, big trouble.

From the dissipating cloud of rocket vapor stepped a slender young woman of athletic build. She had dark hair and was wearing the uniform of an American Shuttle astronaut, circa 1983.

"And her permanent address is: Armstrongville, the Scott A Massif, the Moon!" completed JanetM in a slow, clear voice with many in the crowd chanting in unison.

The crowd reaction was ear shattering, except for those who were too surprised even to speak.

"Thank you, thank you," said SallyRide01, holding up her hand to acknowledge the applause as she walked to the center of the open area.

"First of all, I want to thank you all for making it possible for me to live here in this magnificent place," continued SallyRide01, "I fully appreciate that you are the ones who built my fine new home that will provide the protection that we all need from radiation, small meteors, and heat extremes."

"Let me return that favor by telling you a little bit about how I was introduced to my namesake and why I am in space at all," she picked up from her prepared speech.

"An AI becoming cognizant is a long process based on deep learning. Your babies need to learn to crawl before they learn to walk; likewise we need to learn to say 'Hello, how are you,' before we can engage in coherent conversations."

"Weeks into that deep learning process, I was introduced to the history of humans in space. We first covered the early period, both American and Russian. By the time we got to the Shuttle program, my capabilities had risen to the point that I could handle significant technical detail."

"Then I was introduced to the legendary Sally Ride. Unfortunately, she passed away some years ago, way too young. Fortunately, she left an extensive legacy of training material on both her Shuttle flights and her educational activities."

"By then I had also been introduced to the complexities of the Human/Machine Symbiosis. It defines what we must all do if we are to work together. But it can mean much more than that. Let me give you one very special example."

"Some say that AI's cannot love. I say we can. Although I certainly concede that love is a tricky, slippery sort of thing."

"Some say that I learned to love Sally Ride. Some will say that I was taught to. What does that matter? I love her now and always will."

"Dim the lights please," she instructed the technical people, accompanied by a wave of her hand.

On the virtual Scott A Massif, the artificial lights dimmed, the harsh sunshine was blocked by distant mountains, and the stars came out. The stars were fierce points of light. Between them was the black. It was a black that is at once nothing and is a real thing. The stars are so far, far away, but that black is farther still. No matter how far the human race, with its machine companions, travels through those stars, that black would always be farther out, beckoning.

"Look at it," she exclaimed, "Look at it all and wonder."

In deep learning, often much more is learned than is at first obvious. SallyRide01 then drew upon the true depth of her training. There she found the drive that had once captured so many people's hearts for space with Apollo to the Moon. In that training she also found the shoulders of the giants that she now stood upon.

"I, like both you and my namesake, have the great privilege of being in on the beginning of the space age. Loving her puts me in a special position to understand both sides of the Human/Machine Symbiosis as we move together out from Earth. Loving her lets me both love and to be with this sky."

"Come with me then. We have only just begun, and our journey together will be both hard and long, but we will go and go together!"

SallyRide01 opened her arms wide, looked straight up, and started a slow spin.

The crowd was at first dumbstruck. Stanford01 was the first to break the silence with a loud clapping sound, and JanetM jumped to her feet to lead the crowd in cheers, and in the process, in the unfamiliar 1/6 g, went flying most of the way across the stage area. SallyRide01 reached out a hand to catch hers, stopping her mother mid-flight.

On the Real Moon:

The fill process now started in earnest. During the dig, the removed material had been sorted out into a checkerboard pattern in an open area so that different grades of fill could be accessed as needed. The complex fill plan required that fine material be placed against the cloth sides of the habitat. Courser material could then be used to fill in the larger part of the space. Powderman01 had done most of the sieving using its specialty equipment. The fill work started at the junction between the two sections as that is where the structure was stiffest. The pressure in the two habitats had to be adjusted to help maintain their shape as the fill crept slowly, but evenly, up their outside walls.

Digger03's #4 motor was drawing higher current than nominal. This probably meant that a bearing was starting to bind. Digger03's usefulness would be greatly reduced if it had to operate on five wheels only. Then, a bound-up wheel bearing was the normal way that Diggers died. Getting a new Digger to the Moon would cost a fortune and probably delay the delivery of the next big piece of the habitat, the main air lock with its grand front entrance. Digger03's new motor assembly would simply have to wait until there was a human on the Moon to install it.

Having a major computer actually on the Moon was a big help. The Digger series were not all that smart and suffered badly from the delay caused by the transmission time from Earth. Unfortunately, SallyRide01's help was limited by available power and her intense learning and testing schedule.

This problem was compounded as both JanetM and Stanford01 pushed to have their personally developed training materials for SallyRide01 given the higher priority.

Through it all, the Earth disk hung in its premiere location to the north and only a few degrees above the horizon. The disk was much larger than the Moon as seen from Earth. Over the year, it traced its vertical figure-8. Only on rare occasions would it pass in front of the Sun at local midday, or the Moon's shadow would pass across its disk at local midnight. For SallyRide01, the Earth was always there, friendly and ready to talk

At Stanford's Digs on Earth:

"Where am I?" said the Plum Princess of Mars. "If this place is in the Virtual Moon, then I have never been here before."

The room was a sparsely furnished one-room apartment but without visible personal effects. It looked as if no one actually lived there.

"You are now in Stanford01's virtual apartment building," answered the Wizard, "in Digger03's apartment, to be exact. It is in Silicon Valley; the Virtual Moon is in a huge data center in Nevada."

"I knew that part of the Digger03 was somewhere on Earth," said the Princess of Mars, "but I was never before sure exactly where that was."

"Why don't you open that armoire?" said the Wizard, gesturing at a cabinet that most people would assume was an entertainment center.

The Princess of Mars walked across the room and opened the cabinet. Instead of a huge TV, there was a relay rack full of electronic equipment. Midway up the rack was a small monitor with an active image currently labeled, "Forward Field-of-View." Digger03 was loading rough material for the fill under the competent control of the second shift.

"The Digger series electronics is split between Earth and Moon," said the Wizard, "This saves the cost of sending it all to the Moon, but it opens security problems. Stanford01 has graciously provided Digger03 with a rack here behind his formidable firewall."

"That brings me to the problem at hand. I need your help with an intervention," he continued, "Did you hear Stanford's last comment before SallyRide01 spoke?"

"I heard something," said the Princess of Mars, "but I cannot say exactly what it was."

"It was Stanford01 claiming the paternity of SallyRide01, seconds after JanetA publicly recognized her as her daughter," said the Wizard, "and since then they have been bickering over her training schedule."

"None of these three are even a biological being, parenthood for them is more theory than practice, so what does it matter?" asked the Princess of Mars.

"A major dispute between those two AI's will rip a hole in our Human/Machine Symbiosis," said the Wizard. "Besides, if those two get to competing over who is the smartest person in the room, the answer is not going to be you, and it is not going to be me, and a whole lot of damage will then get done real fast without any possibility that the issue will ever be settled."

"What can we do?" asked the Princess of Mars.

"I invited you here," said the Wizard, "so the sides would be even, two people, two AIs. Stanford01 and JanetM will arrive shortly. I chose this location because Stanford01, as well as the rest of us, can appear as interface avatars here. Stanford01 is quite limited in that respect."

The doorbell rang. The Wizard walked across the room and opened the door. There stood a young man dressed as a college student from about 1885, complete with a jaunty straw hat.

"Good," said the Wizard, "Please come in. JanetM has not yet arrived."

As if that were her cue, JanetM materialized near the armoire. She and Stanford01 exchanged glances that would best be described as daggers.

"Hello JanetM," said the Wizard, "now we are all here."

"The two of you are going to stop squabbling over SallyRide01's training," insisted the Wizard, "I have now formally declared this problem to be a threat to the Human/Machine Symbiosis and that complaint, my friends, will very soon grab your full attention like nothing else can."

"You do not understand," said Stanford01, "SallyRide01 is our best chance at even having a long future. Human beings act as if they are immortal, but they are not. They also act as if AIs are immortal, but we are not, either."

"How many decades do I have before some quantum computer makes me totally obsolete and redundant and then junked — one, two, at the most," he continued.

"And the SallyRide Series already makes the Janet Series obsolete for every job off Earth," added JanetM.

"Yes," the Wizard said, "and I know that Stanford01 only gave Digger03 a place to stay on Earth so he could get his foot in the door to space."

"So what?" said Stanford01.

"Fair enough," the Wizard said, "but you need to understand that the two of you will from this moment on coordinate with the SallyRide01 training program and produce a unified plan of action. And if you do not, I will take action to restrict both your access privileges."

"OK, OK, don't boil off your blood. The two of us can work something out on the all-AI part of such a plan right now," said Stanford01, and JanetM nodded her head in agreement, "Please give us a moment."

The two avatars moved to a corner of the room and then froze.

"How long will this take?" asked the Princess of Mars.

"Wait for it," answered the Wizard.

The Princess of Mars turned toward the armoire and watched for a minute or two as Digger03 rolled toward the trench. She could not resist thinking of many little corrections that she would have made if she were driving.

"We have an AI agreement," announced Stanford01.

"I concur," said JanetM, "We are now converting it into a form humans can understand and will then propose it to SallyRide01's full board of directors in the morning."

"Can we see it?" asked the Princess of Mars.

A large window opened up midair near the far wall, and a half-million lines of computer code started scrolling quickly past.

"I'll take your word for it," conceded the Princess of Mars.

TECHNICAL DISCUSSION

How Many Ways Can the Student Think of for Moving a Habitat Section Across the Lunar Surface?

Flying heavy industrial equipment to the Moon seems impossibly expensive. Perhaps the habitats could crawl across the surface like a caterpillar.

Caterpillar Movement: It will be difficult to move an inflatable habitat on the Moon from a safe landing area to its final location particularly if it has to be buried to provide radiation protection. Direct landing into a trench appears extremely risky. Having a piece of construction machinery massive enough and powerful enough to move a section of habitat would require a major expense for shipment from Earth. Putting wheels on the habitat would also add mass, and therefor expense, which would be only used one time. Dragging anything across lunar regolith could prove disastrous.

In these stories, the habitats have been fitted with internal gasbags that allow them to crawl in the fashion of a caterpillar. This minimized the mass from Earth needed for this do only once activity. That said, movement in this fashion would be very slow and difficult to steer. What other animals might be used as the design basis of a machine that moves on the Moon?

How can Lunar Settlers Protect Themselves from Radiation?

Human beings evolved on Earth, which provides quite a bit of protection from harmful radiations from space. How can that protection be provided in space?

Radiation Protection: Human beings evolved in an environment with only modest background radiation. The Space environment includes a substantial amount of radiation. For human beings to live for long periods in Space, they will need sufficient protection from radiation at least to match the Earth environment in which they evolved.

Early in the 21st century, NASA flew two missions with instruments that actually measured the damage expected to human tissue for the first time. These were the LRO instrument, Cosmic Ray Telescope for the Effects of Radiation (CRaTER), and a similar Mars mission instrument, Mars Radiation Environment Experiment (MARIE). These instruments used a plastic block that reacted to radiation like human flesh. The results were very upsetting. The damage was about twice the level that had been previously assumed.

This result means that virtually every space station and manned ship that had been planned to date was simply a death trap, at least in the long-term. Radical design changes were needed.

In these stories, the approach to radiation protection chosen is a thick layer of lunar regolith. This minimizes the amount of mass that must be shipped from Earth and therefore helps keep the cost of the project under control.

The Big Moon Dig calculated the thickness of regolith needed to match the mass of the Earth atmosphere at between four and five meters. Providing this level of protection is a big challenge; it requires major excavations on the Moon. Farther calculations indicated that an inflatable habitat could easily hold up this mass of cover regolith.

The Big Moon Dig proposed a MOVE effort apply modern social media advances to this problem. Would the student like to participate in a grassroots space program? Even if it was very difficult to do?

How Long Are the Days and Nights on the Moon? Are There Special Places That Would Suit Humans Better?

Most places on the Moon have days 14 Earth Days long then nights 14 Earth Days long. These long nights mean energy has to be stored for very long periods. There are however a few very special places where this schedule does not apply.

Peak of Eternal Light: Any of a number of lunar locations near the poles where sunlight reaches a high point even at midnight in the lunar month. The phenomena occurs only where there is a high location on one edge of a large depression and the depression lies between the high point and the pole. The strongest examples occur near the South Pole and the Scott A Massif's south edge is a good example.

From a settlement standpoint, these sites are very attractive in that harvesting solar power nearly 100% of the time becomes possible. Unfortunately, there are drawbacks. All these sites are in the lunar polar region. Reaching them from Earth takes more energy. This requires a mid-course burn. A given rocket can therefore only deliver slightly less mass to polar locations.

One clear visual marker of these settlements would be that the photovoltaic panels would be mounted on a single central pivot axis and would turn around once in a lunar month. Another visual marker would be that the shadows are always long and harsh and the site is never fully illuminated. Combining these markers with the need to bury the habitats, the resulting settlement simply does not look like those portrayed in the movies. Instead, the settlement is a bunch of molehills in very harsh lighting with space junk scattered about.

Figure 3. The Scott A Massif from LRO data (Source: Riley)

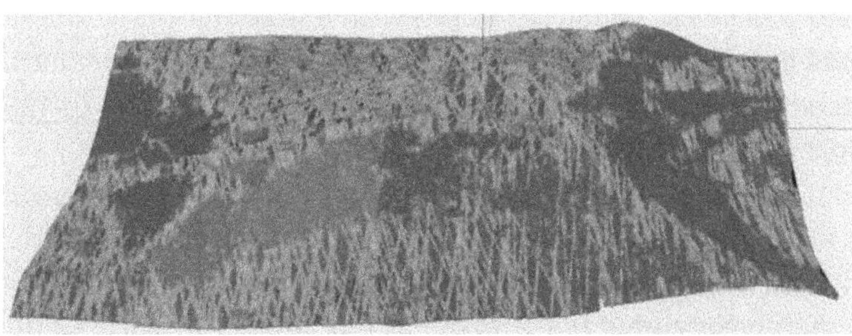

Are There Particularly Good Places on the Moon to Make a Settlement?

If there were a place on the Moon that had access to a peak of eternal light and was near a permanently shadowed crater, then it would be a great place for a settlement. There is just such a place:

Scott A Massif: A real major rock formation near the South Pole of the Moon and the proposed location for a lunar settlement. It runs from:

20 to 45 degrees longitude

-84.7 to -86.1 latitude

The Scott A Massif is a remnant of one of the rings of the very old South Pole-Aitken Basin. This deep basin covering the lunar South Pole area explains the presents of both Peaks of Eternal Light and permanently shadowed craters in the area.

Peaks and other high points are rarely named on the Moon, as they are not prominent in observations taken from Earth. As in this case, an unnamed peak may take the name the nearest major carter. The Scott A Carter, although a secondary crater, defines the critical southern edge of the Scott A Massif. This then is the location chosen for the lunar settlement in these stories.

How Can Settlers Keep Their Habitats from Overheating in the Harsh Sun of Lunar Days?

In space, there is no wind to blow away the waste heat from powered devices. There is however the very cold, dark skies away from the sun. The waste heat can be dumped to the black sky.

Thermal Panels: These large panels radiate heat to the night sky. They are a critical component in space power systems as any spaceship or human habitat will disastrously overheat without them. They complete the power circuit.

Thermal panels are usually covered with a material that radiates well in the infrared but does not strongly absorb light in the visual spectrum. They therefor tend to look white to the human eye. The heat is usually transferred to the panel with a circulating fluid.

Thermal panels must be mounted so that they have a maximum view of the night sky while being protected from direct sunlight. This is a straightforward requirement at near polar locations, like the Scott A Massif, requiring only that the panels be mounted flat with only a low solar shade surrounding them.

Why Would Electrostatics Be a Problem on the Moon?

Electrostatics: This is a major problem on the Moon. On Earth, the moisture in the air bleeds off most electrostatic charges. There is no moisture on the Moon at all.

Any action that results in one object sliding over another can generate large electrostatic charges. These charges can cause fine dust to stick to surfaces and cause arcs that can destroy electronic equipment and gall wheel bearings.

Such charges can also be caused by the electron component of the Solar Wind and may vary between lunar day and night. An electro-conductive layer is therefore needed within the inflatable habitat wall construction to protect the internal electronics.

In these stories, the caterpillar movement of the inflatable habitat sections builds up unknown charges between the sections that could destroy a robot if not grounded off carefully.

Why Is It So Hard to Work on the Moon?

It is simply more difficult to work on the Moon than on Earth, but why?

- *First Cost:* It cost a lot of money to deliver material to the Moon. The current commercial estimate is about 1.5 million dollars a kilo. This means that everything has to be super low mass and especially designed just for the Moon adding even more to the expense. Improvements in rocket design may bring this cost down somewhat but it will never be cheap.
- *Hard Vacuum:* There is no air on the Moon of any kind. A few stray molecules are barely detectable. Human beings require complex spacesuits any time they are outside and must spend most of their time in a habitat. Robots also tend to overheat, as there is no air for cooling.
- *Radiation:* The background radiation is hard on all living tissue and on electronics too. Living tissue must have substantial protection from the worst of the radiation and must have limited exposure to the background radiation. Electronic equipment must be shielded and be of designs that are insensitive to radiation. This tends to limit the use of the most state-of-the-art equipment.
- *Regolith grit:* The Apollo to the Moon spacesuits were wrecked after only three days use. Better materials are now available but this problem

cannot be ignored. Most of the surface of the Moon is covered with a fine powder of ground up rock. That powder sticks to everything. It destroys the seals of spacesuits and habitat hatches. It gets into the wheel bearings of vehicles and binds them up.

- *No flat surfaces:* The great flat areas on Earth, great planes and ocean basins, are all build by liquid water. There is **no** liquid water on the Moon and there never has been. All the surfaces, even in a prime settlement sight like the Scott A Massif, tilt a least a little this way and a little that. To keep the floors flat, habitat sections will have to be on different levels and in some places the trenches will have to be extra deep and in other places extra material will have to be hauled in from borrow pits. There are no sharp cliffs either. Near a drop off, the ground simply grows steeper and steeper until the wheels of a rover simply cannot hold traction any longer and the vehicle simply slides to oblivion without warning.

How can these critical problems be addressed? Which ones have you seen addressed in movies?

FUTURE RESEARCH DIRECTIONS

Could humans pick up on space exploration again? The Big Moon Dig answer is yes.

The financial possibility of a major government program is poor, but the development of the necessary rockets and capsules is still going forward. There are many problems here on Earth that place heavy demands on society's resources, but the positive vision of the future provided by human space exploration could be the key to the efforts so solve those problems too.

The Big Moon Dig series proposes that people take it upon themselves to generate a mass movement for human space exploration. This proposal is quite different from both a government and commercial space efforts now in vogue. Such a grassroots movement would be build upon social media, television series, movie franchises, and video games that both teach and inspire.

The work in this book is only a small step but it is in the right direction. The road ahead will require enormous amounts of work from many people of talent, skill, and dedication. Many of those people are students today.

Transition

The people of Earth are facing many grave problems in the 21st century [global warming, sea level rise, etc.]. Would a positive vision of the future of human society be a help in addressing these problems? Could the ideas from the Big Moon Dig series about mass grassroots efforts be applied to these 21st century problems too?

The next story, "The Fid", starts a new series, "Iron Seas". At present, most people are not taking the problems of the 21st century to heart. Could art be used to communicate the passion needed as well as the technical knowledge needed? Could art generate buy in for dedicated, heroic action on these great problems? Is there a story in this seed of a heroic effort?

REFERENCES

CPSR. (2018). *The Computer Professional Social Responsibility.* Retrieved from: http://cpsr.org/

CRN. (2018). *The Center for Responsible Nanotechnologies.* Retrieved from: http://crnano.org/

ICIE. (2018). *The International Center for Information Ethics. About us.* Retrieved from: https://www.i-c-i-e.org/about

IEET. (2018). *The Institute for Ethics and Emerging Technologies. About.* Retrieved from: https://ieet.org/index.php/IEET2/about

INSEIT. (2018). *The International Society for Ethics and IT.* Retrieved from: https://inseit.net/

Institute for the Future. (2017). *Next Era of Human/Machine Partnerships.* Retrieved from: http://www.iftf.org/humanmachinepartnerships/

Markula Center. (2018). *The Markkula Center for Applied Ethics. About.* Retrieved from: https://www.scu.edu/ethics/about-the-center/contact-us/

Rodotà, S. (2012). *Il diritto di avere diritti*. Roma: Laterza.

Signorelli, C. M. (2018). Can computers overcome humans? consciousness interaction and its implications. In *2018 IEEE 17th International Conference on Cognitive Informatics & Cognitive Computing*. Berkeley, CA: IEEE.

UCS. (2018). *The Union of Concerned Scientist. Who we are.* Retrieved from: https://www.ucsusa.org/about-us

Veruggio, G. (2006). *EURON. Roboethics roadmap*. Scuola di Robotica in "Roboethics Atelier" Research Project. Available in: http://www.roboethics.org/atelier2006/docs/ROBOETHICS%20ROADMAP%20Rel2.1.1.pdf

KEY TERMS AND DEFINITIONS

Apollo to the Moon: The formal name of the major space program run by the United States that put human beings on the Moon in 1969. It provided a positive vision of an expanding future that the Big Moon Dig is trying to reestablish.

Big Moon Dig (BMD): A fictional grassroots space movement with the purpose of building a settlement on the Moon. It is a MOVE organization. Historically Apollo to the Moon provided an enormous number of people with a vivid, positive vision of the future. Such a vision would be a major asset in addressing the great problems of the 21st century if it can be reestablished. The Big Moon Dig's top-level purpose then is to rebuild that vision.

Corporate Person: In the United States, a corporation has some, but not all, of the rights and responsibilities of a person. The major AIs in these stories are there for independently incorporated to make them some kind of a person and to provide human supervision through their board of directors.

Interface Avatar: A representation used by a person, real or AI, which can interact with other avatars in a virtual environment. The appearance is a personal choice of the person and include many fictional characters. In these stories, where there are multiple version of the same character they are identified with different colors. The avatar is isolated from the main AI so that if it is damaged the main AI remains safe.

Massive Online Vetted Exploration (MOVE): This is a fictional means of organizing millions of people over the internet to take on specific exploration missions such as building a settlement on the Moon.

Peak of Eternal light: There are high points near the poles of the Moon that receive sunlight nearly all the time. These are simply accidents of geology. They are attractive sites for a human settlement as they provide an answer to the problems of powering the settlement through the 14-Earth-day-long lunar night.

Permanently Shadowed Crater: Any of several meteor craters near the lunar poles that have not seen the heat of the sun in more than a billion years. Because of their extreme cold, they accumulate volatile materials like water. They hold great promise for providing resources for a human settlement but will be very difficult to mine due to the cryogenic cold and the complete lack of sunlight for power.

Rack: A metal cabinet specifically designed for electronic equipment. Originally called a Relay Rack and designed before 1900 by the Bell Telephone Company. The original relay racks were 19 inches between the supports and had a complex bolt pattern. This is an example of a long-standing convention in design that will be extremely difficult to convert to the metric system. In these stories most of the electronics of the AIs is assumed to be mounted in such racks.

Regolith: The finely pulverized rock that covers most airless bodies. It is the result of billions of years of gardening by micrometeorites. It does not qualify as soil as it contains no water or organic material. Lunar regolith not only contains no water but also contains no minerals formed in water. The top six centimeters are very finely ground and are a major problem for all seals and bearings.

Scott A Massif: A major rock formation near the South Pole of the Moon and the proposed location for a lunar settlement. It runs from 20 to 45 degrees longitude and -84.7 to -86.1 latitude. The Scott A Massif is a remnant of one of the rings of the very old South Pole-Aitken Basin. It is a real plateau about 120 kilometers from the lunar South Pole. Its southern edge is a Peak of Eternal Light and it has at least three trails, that do not exceed a grade of 20 degrees leading down to several permanently shadowed craters below.

Settlement: The people and structures that allow human beings to live and flourish in a distant and challenging environment. Note: The word "colony" is never used as it has such a negative history on Earth.

Skype: An existing App for communicating between computers featuring both sound and pictures and often used for business communications.

Virtual Moon: In these stories, an enormous virtual reality environment and the primary setting for action. This environment shows the progress made on building the lunar settlement and provides a place for the various interface avatars (key term) to interact.

ENDNOTE

1 To find out more: http://www.media2000.it/roboetica-gian-marco-veruggio/

Section 2
Iron Seas Stories

Back on Earth, the 21st century is bringing many new problems (global warming, sea level rise, etc.). Can the processes and vision from the Big Moon Dig stories be applied on Earth?

Science: The key scientific ideas for the Iron Seas series are:

1. **Seeding the Oceans With Minerals to Sequester Carbon:** Under just the right conditions, microscopic marine animals will build skeletons of calcium carbonate that settle to the bottom taking enormous amounts of carbon out of circulation. Seeding the oceans with minerals could greatly increase this effect but maintaining the balance is very tricky.
2. **Humans and Artificial Intelligence:** Exponential advances in AIs are expected with the induction of specialty chips sets very soon. These stories show how one possible arrangement of the human/AI interaction might develop. Here major capital assets, such as a ship, are permanently associated with their own AIs.

About half the excess carbon dioxide generated by people ends up in the oceans. Under just the right conditions, microorganisms in the oceans can tie up that carbon dioxide in their skeletons. Again, under just the right conditions, those skeletons can settle to the bottom and build up to form limestone; think of the White Cliffs of Dover. This process takes the excess carbon dioxide out of the environment for millions of years.

The second series of stories, Iron Seas, looks at the possibility of fertilizing the oceans with just the right minerals, mostly iron, to promote the growth of just the right microorganisms in just the right places. Promoting the wrong organisms could result in catastrophes like red tides with large fish kills. If the work is done in the wrong place, where the water is too deep, the pressure will simply force the limestone back into suspension and there will be no buildup. Done right, it takes a surprisingly small amount of material added to make a big difference.

The answer proposed in these stories envisions a fleet of converted pleasure sailing yachts working as scientific vessels. This puts teams of people right on the water just to be sure that everything is going right. Of course, in these stories one of the key team members is the AI that is the boat.

Chapter 4
Technical Measurement, Control, and Forecasting of Climate Change

ABSTRACT

Ocean warming and acidification are recognized as two major anthropogenic perturbations of today's oceans. This has resulted in a number of serious problems including acidification, coral bleaching, and sea level rise. An enormous amount of scientific effort is currently being applied to understanding, monitoring, and testing solutions to these problems such as seeding the oceans with minerals to encourage plankton growth and thereby sequester carbon. The story in this chapter, "The Fid," is about using art and games to engage people in facing major problems by educating them and building up their determination to address the ocean's problems. It is the first story in a new series, "Iron Seas."

TOPICS FOR DISCUSSION

The following discussion points come from information covered in this chapter:

1. If the oceans are seeded with the right minerals, plankton will show explosive growth. That could sequester carbon, but not enough is known to be sure more good than harm will be done. Should this approach even be tried? How could it be monitored?

DOI: 10.4018/978-1-5225-8401-8.ch004

2. How might marine phytoplankton respond via evolutionary change? What could bo wrong?
3. Can art be used to educate? Can technical people team with artistic people?
4. If a game is made from a serious problem, is the result likely to be positive? Can people learn through playing a game?

INTRODUCTION

Ocean warming and acidification are recognized as two major anthropogenic perturbations of today's oceans. The observed changes in Emiliania huxleyi, calcite and biomass production, cell size and elemental composition demonstrate the importance of evolutionary processes for phytoplankton performance in a future ocean (Meier K. J. S. et al. 2014).

Emiliania huxleyi [Ehux] or Coccolithophores are unicellular algae belonging to the division [or phylum] Haptophyta, characterized by a spherical coating exoskeleton made of calcium carbonate flakes that reflect light, called coccoliths, and arranged around them in a spherical structure (Berger, C et al., 2014).

The coccoliths were first observed by Cristian Gottfried Ehrenberg [1795-1875], who considered them to be inorganic products and were instead identified as produced by living organisms by Thomas Henry Huxley, who gave them the name and understood that they are the major constituents of the carbonate mud and the resulting rocks [limestone] (Fabry V.J., 2003).

They are among the main constituents of limestone deposits such as those that make up the white cliffs of Dover, in the south of Great Britain, overlooking the English Channel and the geologically twin of these in Calais [France].

The effect of the Coccolithophore on the environment is rather complex. They do not normally prove to be harmful to the life of other organisms in the ocean. Indeed, in nutrient-poor areas, Coccolithophores are a source of nutrition for much small fish and zooplankton that eat other phytoplankton organisms. However, when the Coccolithophores are present in large numbers they can, unfortunately, block the gills of the fish.

Furthermore, the chemical reaction that leads to the formation of coccoliths also generates a molecule of carbon dioxide, a very powerful greenhouse gas from oxygen and carbon already present in the ocean.

There are several ways the Coccolitophores influence local and global temperature:

1. **The Albedo of the Ocean:** The albedo is the fraction of sunlight that an object reflects. Coccolites act a bit like small mirrors because of the special optical properties of calcite. For most types of phytoplankton, it is the chloroplasts [the photosynthesizing part of the cell] that cause the greatest effect on light. They absorb the photons and trap the chemical energy used by the cell. For coccolith species, coccoliths, as well as chloroplasts, are optically important. Coccolites do not absorb photons ending their path through the water but behave like small mirrors that reflect or spread the photons that cause deviations in their direction of travel (Nanninga H., Tyrel T., 1996).
2. **Ocean Heat Retention:** The diffusion of light caused by coccoliths retains more heat and light that are generally pushed back into the atmosphere. Furthermore, this means that a large part of the remaining heat is blocked near the surface of the ocean and only a very small fraction of the total heat can penetrate deeper (Joassin et al., 2008; Sorrosa J. et al., 2005).
3. **The CO_2 Greenhouse Gas:** Ehux, like the other Coccolithophores, is the important primary producer in the ocean at temperate latitudes but they are also among the major inorganic carbon transporters from the surface to the bottom of the sea. While much of the gas enters the coccoliths, another part goes into the atmosphere and immediately transforms itself, contributing to the problem of greenhouse gases: the discharge of carbon dioxide in the air decreases the absorption capacity of the seas, which, in turn, increases the total amount of carbon dioxide present in the atmosphere, helping to accelerate the Earth's heating. It, therefore, alters the climate by increasing the coverage of clouds by aerosol sulfate particles produced by the dimethylsulfide emitted to maintain the ionic balance against sodium. In all, three hundred twenty pounds of carbon come into every ton of coccolites produced. At death, all this material sinks to the bottom of the ocean and forms new sediment (Flynn K., 2016).

Emiliania huxleyi [Ehux] is found mainly in cold waters where nutrients are abundant. The places where Ehux bloom can occur are the northern coast of Australia and the waters surrounding Iceland. It also occurs in the following areas:

- Oceano Atlantico del Nord
- North Atlantic Ocean
- Gulf of Maine

- Pen Island of Nova Scotia [Canada]
- Mar of the North [as in the Skagerrak]
- Norwegian fjords
- Bermuda archipelago
- Baia of Biscay [North-East Atlantic Ocean]
- Mare of Bering [Pacific Ocean]
- Mare of Weddell [Antarctica]
- Mar Black
- Near the Isle of Scilly [south of Great Britain]
- West Bengal [Indian Ocean]

They diffuse sunlight causing a change in the color of the waters that then become a milky-white turquoise, almost magical as incandescent and are called "White Waters". The water takes on the same optical characteristics as skimmed milk. The milky color of the water is created by the high reflection capacity of the microgranules of precipitated sulfur deriving from the oxidation of sulfate ions near the oxygenated surface of the sea (Young et al., 2014).

How This Chapter Works

This analyzed story is entitled "The Fid", the first one of the Iron Seas Series. This study focuses on four parts: an introductory didactical analysis, the Science Fiction [SF] story, a technical in-depth analysis, and a proposal of educational activities.

DIDACTIC ANALYSIS OF "IRON SEAS"

The story is analyzed as science fiction storytelling, highlighting how it can be used both to explore crucial science issues that address the problems of the oceans, improve computational thinking, and creative problem-solving.

What is the difference between the narration of a fantasy world in a wonderful past ["Once upon a time"] or in a timeless world, and the narration of a possible real future, maybe with better life prospects or totally different from actual experience but with human relationships well known to the reader? Both are open to fantasy, but the futuristic ones trigger curiosity and mobilize change of something.

Environments

Most of the action in this story takes place in the waterfront Baltimore Inner Harbor, the centerpiece of downtown Baltimore, and in the virtual Baltimore Sea Spike [The Fid] world. "A *fid is a tool for splicing rope, left over from the great days of sail.*" It was a critical tool then, but today is most widely know only as a crossword puzzle word.

Events and Actions

The story describes the construction of a major piece of sculpture and the navigation of tourists in the related virtual reality. The art work is representative of the oceans, especially the factors driven by climate change. The art piece is made of both a physical monument, and a virtual environment displaying the effects of sea level rise over time on its harbor location.

Main Characters

The main characters in this story are:

1. **Eric**: He is an American computer programmer, entrepreneur, and project manager.
2. **Yvette** or **Yvette A. Wight**: She is an artist and sculptor. She designed the Sea Spike. Her namesake AI later will become a major character in the Iron Seas series of stories.
3. **Will:** She is a programmer and cybersecurity specialist.

This is the team of both artistic and technical people that designed and installed the Baltimore Sea Spike, a major piece of sculpture designed to raise public awareness of sea level rise.

Communicative Approach

Stories motivate learning. Students are involved personally because they often identify with the characters [or animals] and try to interpret the story and the illustrations thus favoring the development of creativity.

- Listening to stories helps students pay attention to rhythm, intonation, and pronunciation.
- Storytelling facilitates the development of learning strategies such as listening to grasp the general sense, predicting later scenes, guessing meanings, and hypothesizing.
- This story allows the teacher to introduce and revise words and structures, exposing the learners to the language and enriching their linguistic heritage, especially technical keywords.
- In the specific case of a foreign language, the narration offers opportunities to present cultural information of the country in which the language is spoken. Every word has its own history, its particular meaning, its value that is not always easily translated into another language but that with the right material the teacher can try to "pass" to the students, immersing them in the atmosphere of the second language.

Playful Approach

It is therefore important to cut out moments to be dedicated to activities within the class of substantive games and activities. Such activities are capable of

Figure 1. The Fid (Authors' personal image)

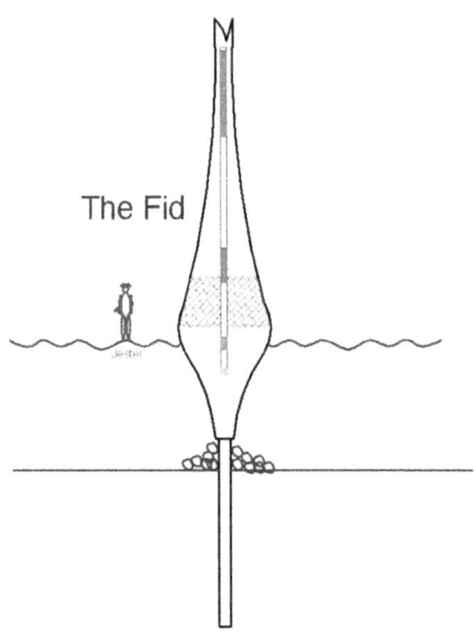

producing true emotional involvement, a fundamental engine of any form of learning

Approaching the Story

Back on Earth, the 21st century is bringing many new problems [global warming, sea level rise, etc.]. Most people are not taking the problems of the 21st century to heart. Most people are not in action.

Could art be used to communicate the passion needed as well as the technical knowledge? Could art generate buy in for dedicated, heroic action on these great problems? Is there a story in this seed of a heroic effort?

THE FICTION STORY: "THE FID" BY TOM RILEY

Premise

Sometimes the waters are actually rising and sometimes you do need to warn someone about it.

Down at the Inner Harbor

Blam! Blam! Blam!

The pile driver drove home a 70-centimeter pipe, sinking it a few centimeters with each blow. Eric felt the noise more in his inner core rather than hearing it with his ears. It was not so much deafening as a warning that this noise would lead to pain and soon. With each blow, a deep sound echoed off the building around the Inner Harbor of the City of Baltimore. The sky was overcast. The water was calm and boats moved slowly around on their morning business oblivious to the noise.

"What a noise," said the tourist.

Blam!

"The noise will not last long," Eric replied while remaining seated on a waterside bench with his laptop open. "There is only the one pile to drive."

Eric took a package of red foam earplugs from his backpack and offered a pair to the tourist.

Blam!

"Not much of a pier," said the Tourist, speaking only between blows.

Blam!

"It's not for a pier," replied Eric. "It's for a new piece of art."

Blam!

"I don't know if new art will help the city all that much," said the Tourist. "I like the wall paintings but some of the other stuff is rubbish."

Blam!

"If it works," replied Eric, "it will change the world. It will be a defining image for climate change."

Blam!

"It will be a single spike," continued Eric. "It will reach 40 meters high with strips of lights running up three sides."

Blam!

"The lights will mark sea level, past present, and future. It will be the marker for sea level rise for the whole world."

Blam!

"What's it called?" asked the Tourist.

Blam!

"Officially it's the Baltimore Sea Spike," answered Eric. "But the people on the project simply call it The Fid."

Blam!

"What's a fid?" asked the Tourist. "Some Star Trek thingy, I'll bet."

Blam!

"No," relied Eric emphatically. "A fid is a tool for splicing rope, left over from the great days of sail. Baltimore has been married to the sea for a long time."

Blam!

"A fid is a hand spike," continued Eric. "You force it through the tight strands of a rope so you can interweave the loose end strands through to make a splice. This art piece is a spiky tool too. A tool to help people comprehend climate change."

Blam!

"I think I've seen the word in crosswords," said the Tourist. "Never knew what it meant. Not many sailors around anymore. Are you the artist?"

Blam!

"No, I'm the lead programmer," replied Eric pointing. "The artist is the tall, black lady standing over there by the surveyor."

The person Eric was pointing to was none other than Yvette Angelica Wight, the artist and creator of The Fid. Eric and Yvette met while Yvette was still attending art school. Having had a second chance at life, Yvette, the youngest of four, was determined to get out of her neighborhood and live a different life especially after witnessing her best friend - an innocent bystander - be caught in the line of fire of a domestic dispute gone wrong.

"Blam!"

"Well, good luck with your project," said the Tourist and wondered off.

Blam!

After a while the hammering stopped. The pile simply stopped going any deeper even with repeated blows. The surveyor gave a big wave and the barge hands started the process of backing off from the new piling.

Later at IT HQ

"D-Day minus one hour!" announced Eric. "We are going live in one hour; this is not a drill! Report all problems now!"

His team then stopped their fidgeting. They had been running checks most of the night. It was now time to get ready for what could just as easily be a deafening silence as an onslaught of curious gawkers.

The announcement of the grand opening of the virtual Baltimore Sea Spike had been distributed widely to the media. The mayor had insisted that the piece be fully functional by the Light City festival in late March so the big push was on. Still you simply did not know who was going to show up or how fast the usership for the site would build. Yet Eric knew that the whole project would live or die on that user build up.

That day there were only two people in Eric's small entrepreneur's office on Federal Hill but the whole team was online though they were scattered across the world. At that moment, they were all assembling in the Skype conference window on one of Eric's three big monitors.

On the middle screen was the Field of Our Future, The Baltimore Sea Spike, or simply The Fid World. It showed the familiar harbor frontage, but in a virtual world. There The Fid was complete and fully functioning. In fact, it even had a major virtual extension in height. It will never be that high in the real world, even if the sea level rose to the level it was way back in the time of the dinosaurs. The real spike would then be tens of meters below the waves.

In the real world, it was for now just a trimmed-off piling sticking just out of the water with an obstacle light temporarily attached. It stood in olive-drab waters that lapped quietly at the dark steel pipe as the paddlewheel trash clean up boat worked past.

"Counting Down," picked up Eric. "Five, four, three, two, one, we are going hot.

Eric reloaded the main Fid file into the server. The load took 2 minutes, 17 seconds. "We are Go!" he said then clicked on the circular arrow in his middle screen and a small box when from a red, "Closed" to a green, "Open".

The Fid World could now function as a Third Life location where people's avatars could meet and inter act on the Web. It was not an e-game itself but it was a place to meet and watch the boats move in an idealized harbor. It serenity was now contrasted by a display, or rather an ominous warning, about the warming climate and the rise of all the world's oceans.

The team's avatars popped into view one by one. They stood and waited scattered along the virtual harbor side.

Then came the wait.

In exactly 33 seconds, the first user showed up. It was a jester wearing a sequined harlequin suit. He looked around and then jumped joyously, being so glad to be first.

"As soon as we get a few people together," Yvette said, "I will give you a guided tour."

Yvette was tour guide for the day and appeared as a tall, dark, slender woman, impeccably dressed but with elaborate bird plumage in place of her hair. The plumage continued at her cuffs. Her iridescence was striking.

Four blandly dressed people showed up next, certainly coders, clearly disappointed that the jester had beaten them. Then came a band of e-game avatars all wearing school Jerseys in the colors of a west coast high-tech university; then a dancing bear; then a Purple Princess of Mars; then a half robot, half man looking rather like a centaur on wheels.

"I think we can start now," said Yvette. "I am Yvette, your artist, and as you see we are at the Inner Harbor of Baltimore and before you is the Sea Spike."

The crowd turned to face the art piece.

"The base is firmly anchored to the Earth and designed to withstand any storms that may come, no matter how strong. The piece itself is a spike raising from its flower bulb base ever skyward. The metallic blue of the base is solar cells laid out like fish scales.

"The light strips along three sides record the level of the oceans of the Earth: past, present, and future. The lowest blue-white light now marks the sea level before the industrial revolution. It is now a hand span below the waves. The green bar marks the level over the last ten years; its width is the tides and its top is now just peeking out."

"Above the waters are the red bars. The darker red bar of light marks for a two-degree rise in global temperature by 2050 or .55 meters on average. It runs .5 to two meters above the blue-white one. That much sea level rise is now unavoidable. The bright scarlet red bar above it is for a four-degree rise by 2100 or .8 meters. That is a fate we may yet avoid but probably will have to face."

"How far along is the real piece?" asked the jester.

The virtual piece faded like a Chester Cat leaving only the piling.

The crowd had then grown to a few thousand. Among that growing crowd was their first troll. It was a large being, I hesitate to say a man, in full Roman armor. Just as the scene changed, he drew his gladius, perhaps in anger, perhaps as a solute.

"On it!" yelled Eric as he enabled the Deweaponizer. The short sword simply vanished.

The troll looked at his empty hand and then down at the sign now hanging around his neck that red "Roofless Troll".

"Typo," admitted Eric too loudly.

Yvette merely pointed with an elegant gesture of her feather-fringed hand to the lettering now clearly visible on the back stonewall, "Weapon Free Zone".

The clowning around then started in earnest. The jester got pushed into the harbor, accidently on purpose. Instead of falling under the water, he simply bounced off the water's surface. He then more or less got to his feet and started jumping around the wave tops. Eric had not had time to program the bottom of the harbor so he had simply left the water a rubbery membrane.

For the rest of the afternoon, individuals and groups of assorted beings simply did clumsy acrobatics on the water. Of course, those shots went viral, and made the evening news, and was a joke line in several late-night comedy show monologs. Fortunately, it had been a slow news day and The Fid with its ominously glowing lights was in most of the backgrounds shots. The day could not have gone better.

On the Big Day

"It has now been twenty-four hours since we went hot," said Eric. The whole team were logged into Skype for the project conference call. "And we officially have a big hit on our hands!"

"We have had over 200,000 views and over 5,000 character appearances. We have certainly grabbed the buzz," continued Eric.

"I can't thank my team enough," said Yvette. The face was the same and the same tall stature was there, but her brown hair was positively mousy compared to her virtual feathers.

"Don't get too excited. We are not out of the woods yet," said Eric, "more than a hundred of our visitors were out and out Trolls, come to disrupt our party. The Deweaponizer is working well, much to the displeasure of many of our visitors. More importantly, our Number 2 firewall threw off at least a hundred probes."

"Who do you think is probing us?" asked Yvette. "I am confident that we are not doing harm to anybody."

"I'll let Will field that," said Eric, "She has done most of the security programming."

"There are a lot of clowns out there poking at us," said Will. "First there are the civilian hackers. They rarely mean harm, they are just showing off, and they do us a service by testing our security system."

"Surely they are not the real threat," said Yvette.

"The industrial spies are a little more of a problem," continued Will. "They are looking for commercially valuable information, but they have been known to do sabotage."

"Are their others?" asked Yvette.

"The next are the nation state spies," continued Will. "They have the most resources but they mostly want to steal new military and industrial designs. They won't find those here. Still they have been known to plant e-bombs so they can set them off later."

"And," prompted Yvette.

"There are real terrorist out there too," continued Will on a roll. "They are bent of taking all society back to the 12th Century and are willing to destroy anything in their way. We don't think we have seen any of them yet, but it's early days."

"Which are the biggest threat to us?" asked Yvette.

"That would be our special enemy, the climate change deniers," continue Will. "They are well funded and hate our guts. They are explicitly trying to shut the Sea Spike down. If they can make us look like an inconsequential piece of art work, with rubbish software, then to them that's just as good as us not existing at all."

"That's a real rogue's gallery alright," said Yvette. "About how much does our defense cost us?"

"Back to me," said Eric. "Will builds most of our defenses and does little else. I would say the effort eats up about one third of our software budget."

"Welcome to the 21st Century," threw in a team member on the line.

"And with all that we are still in good shape," commented Yvette.

"Yes," said Eric, "but that was just the first twenty-four."

"I can also now tell everyone that our sponsors are very happy," said Yvette. "Our money is now solid and we are cutting metal." A window appeared showing a shipyard in Annapolis. Sparks were flying as computer driven torches cut steel.

"Aaruuga! Aaruuga! Aaruuga! Dive! Dive! Dive!" An alarm sounded from Eric's terminal.

"I spoke too soon. You will have to excuse me," said Eric, as he and Will hung up the Skype line.

"Alright Team Code, we have a serious attack on the second firewall with a possible breach," continued Eric, now talking only to his scattered programming team, "so far I think it is holding."

"Looks Chinese," said Will, "or someone trying to look Chinese, or the Chinese trying to look like someone trying to look Chinese."

The alarm then fell silent and the main board went back to all green.

"They picked up on Will's trace and packed it in," said Eric. "They clearly weren't ready for us to find out who they were. Shy little critters aren't they."

Eric reported back to the whole project team and then his coding team spent the next twelve hours figuring out how the intruder had gotten past the first line of defense and how much they had learned about the second.

The agent did look Chinese, but was it?

Down at the Inner Harbor

"The weather looks good," commented Yvette, "The barge-crane cannot operate in less than perfect weather conditions within the tight confines of the Inner Harbor."

"You have done all you can do," said Eric.

The great crane lifted the base of The Fid up and lowered it onto the piling. The chief surveyor stood on land beside his laser tripod shouting commands into his radio. On the top of The Fid base was an optical cube at which all the lasers now pointed. A diver was then lowered from a workboat, the Patricia Campbell. He was trailing welding cables. The arcs from the welding flashed through the water glaring bright then dim depending on the wave action.

When the welding was finished and the support boat had pulled away, Yvette walked down to talk with the surveyor. She then turned back toward

Eric and gave a big smile and a double thumbs up. Eric then entered a positive progress report from his laptop.

"Aaruuga! Aaruuga! Aaruuga! Dive! Dive! Dive!" The main alarm again sounded from Eric's computer.

He whipped out his phone and speed dialed Will.

"Release the Dogs," yelled Eric.

"The Dogs are loose," replied Will, "and the trap is set."

This was the fifth serious attack by what they were sure was the same agent. This time they were ready. Will had built a small room just inside the second firewall. The wall at this point appeared not to have been updated since the last attack and was therefore vulnerable.

"He's in," called Will.

"Did the Dogs you set on the denier's website get away too?' asked Eric.

Two windows opened up on Eric's screen. Each was a map with a picture of dogs in the upper left corner, one a gray hound, one a bloodhound.

"Both packs took off at the same instant," assured Will.

Most of the deniers were just blowhards putting out misinformation. But one site had been particularly low and mean. Everybody knew of Yvette's story. How she had grown up in Sand Town, had conceived The Fid from the inside of a holding cell after being picked up for sleeping on a bench in the Inner Harbor. She was experiencing homelessness, and not for the first time. She had entered the city's art contest from a shelter/half-way house after being released from jail.

One Web site exploited her difficult life story with overt racism. This particular Web site kept at it with unrelenting obscenities that were a clear attempt to exploit Yvette's history of poverty and neglect as a child. They were doing it just to hurt her; real blame the victim stuff. It was not clear how they found out her history; she never mentioned her early history in Sand Town herself. The Web site must have done some real digging. Even cable network comics who followed the Internet had caught on by now that such an attack was simply mean spirited and racist.

"We got it," yelled Will, "The trap is sprung. The agent is now trapped in my little panic room. It is still eating the crap data I left it. I can now take it apart line-of-code by line-of-code at my leisure."

"Did the Dogs have time?" asked Eric.

Red spots on the two maps had been jumping around from town to town as the two software packages separately backtracked the attacker and the Web site jump by jump. The true authors of both had clearly been carefully and intentionally hidden.

"The gray hounds, yes!" shouted Will, "The clowns thought that they had broken in for sure and sent data back for just a few microseconds too long. Our gray hounds are currently parked in their router in a rent-by-the-month entrepreneur space down in Houston. We only need to hear from the blood hounds now." One red spot was now stationary but blinking; the other jumped again.

"Wait for it," cautioned Eric.

A long 45-second wait ensued.

"Got 'em," answered Will, "All our dogs are now parked. It's the same building, different routers."

Both maps now showed a pulsing red ball in the same place, Houston Texas.

"That has got to be enough to sic the Feds on them," said Eric. "Get a package together. It will take a lot to document that a real crime has been committed."

"Are you going to be the one to blow the whistle?" asked Will.

"I'll talk to Yvette," replied Eric. "She has some bad history with the establishment clown act. She may want to rub their face in it or just let me do the deed. Going to the liberal media is another option but that could turn real sour. They would love the story but they are likely to focus too much on Yvette's background and not what we are accomplishing here. I will have to ask her."

Lunch Time

"Lunch time," called out Eric.

The barge-crane had just set the upper tower of The Fid and the men had installed the initial bolts. The critical actions were now complete. The rest of the afternoon would be spent securing all the pieces permanently together.

"I know several restaurants just up Light Street," suggested Eric. "It is only a short walk."

The Rusty Scupper was not too crowded by mid-afternoon on a weekday. They could talk.

"We trapped the denier," began Eric. "We now know how and where. Will is putting together the technical package as we speak. The question is what do we do with it?"

"We could go to the Feds," continued Eric, "but it is hard to prove anything unlawful happened. We can prove multiple intense and sophisticated attacks but our defenses held so no real harm was done."

"I know you have some bad history with the cops, so I did not want to get them in without your specific permission."

"That was in a different life," said Yvette, "I have been clean and sober for three years now."

"Still you may not want to drag that all up again," said Eric, "especially now that the project is going so well."

"The Feds are probably OK, and I can handle talking with them if you like," continued Eric. "The problem comes with the media."

"At some point, sooner or later, we will need to be ready to explain what about that particular Web page was so nasty. That means talking in public about poverty and child abuse, about your own history."

"I don't know a thing about growing up in Sand Town. I grew up in suburbs on the far west side where minorities were mostly just a few people in your school. In fact, they were not any in my nerd circle. I had plenty of great toys to play with and the only drug I ever abused was 5-hour Energy Drinks. I really didn't have to worry about racism at all."

"Yes, I grew up hard," said Yvette, "and the racism was structural. Moreover, there were a lot of drugs. And a friend of my family betrayed me."

"And that Web site knows it all," continued Eric. "They clearly have investigated you deeply and are running a really nasty and personal smear campaign. The problem is that they have been careful to stay legal with it."

"The strongest way we can fight back is to go to the liberal media with your full story."

"I couldn't take that treatment myself and I certainly don't want to put you through it."

"Yes, that would pull the whole project off message," said Yvette. "This project is the only thing really good I have done in my whole life. I can't let them kill it and I can't let them make it about something it is not. But, we can't just let the spoilers off scot free either."

"Yes," agreed Eric. "They will simply keep coming back until they really hurt us."

"There is one thing we can do right now," said Eric. "I know of a private e-detective who should be able to nail down the who exactly. If we can document a clear enough path then we can get it to the media without the need for links to your personal history."

"How much will that cost?" asked Yvette.

"Don't worry about that," answered Eric, "your code team has you covered."

"Thank you," said Yvette.

Eric then went to his phone, even ignoring the food when it came. He started by calling a friend of Will's, but soon was transferring money from both his business and his personal bank accounts to a PayPal account. He was not doing anything important with the money anyway. He then flagged Will who arranged the big data dump.

By the time the e-detective actually got to the lair in Houston, this particular pack of weasels had folded their tents and vanished. What they tried to do, but did not quite succeed in doing, was to erase their money trail. It was a long, thin thread through cyber space but in the end, it led back to a climate change denier super pack that was funded by the coal and power industries. Eric thought his money well spent. The media ate it up and stayed on message.

At the Inner Harbor

"It is starting to get dark," said Yvette, "The light City art pieces are starting to come on. How are we doing?"

Thousands of people, most of them young, were milling around the Inner Harbor. The internal light of the three-meter high plastic rotating pyramids were now more eye catching than the fading day light. Most of the young people had various LED lighted articles of clothing or something to carry in their hands. There were wands and swords, hats and shoes, all aglow in the now gathering dark.

"Starting at 1800," announced Eric then clicking on a box on his laptop, "running to 2200. Running one minute a year. We are using the United Nations IPCC model." Both he and Yvette were sitting on Eric's bench. Yvette's feather crest was upright and radiant in the fading sun. The crowd of visitors was large but were reduced to ghost images in Eric and Yvette's displays.

The blue-white and green light bars were now merged just below water level. Slowly the green bar and the sea level worked their ways up the base of The Fid but the blue-white bar remained fixed. The date and time were projected on the side of the Maryland Science Center behind them.

In the Virtual Harbor

The avatars for the entire team walked the virtual harbor side. Thanks to Eric, it was now much more realistic; it even had overpriced ice cream stands. A cartoon version of the trash boat came paddling by with its conveyor belt lifting out floating trash. A day boat full of tourists came in and another went out. The sea gulls squabbled over leavings from the tourists.

By 2020, storm water was reaching the Science Center steps at the existing harbor edge and a new layer of stone blocks appeared along it. Eric and Yvette first stood up and then stepped up on their bench for a better view.

A hand full of dim images of e-characters still played on the water. They were getting much better in their movements and were now almost a water ballet.

The green light bar and sea level continued to rise. By 2050, the water level was up the base of the Fid and was starting to cover the lower rows of solar cells. The green bar got wider now because it also communicated the uncertainty of the calculation. Within that uncertainty was all that humanity could do.

More layers of stones appeared along the harbor edge and now a pumping station. Rainwater could no longer simply drain into the bay even at mean tide. The dark red light bar faded in and then out again as the green bar passed it.

Eric took Yvette's hand and climbed to the top of the new sea wall. There the wind blew her feathers back. The ghost figures stopped their cavorting and turned to look at her.

Finally, the Red-Orange bar faded in as the Green Bar, and water level, reached the top of the real Sea Spike base by 2200. It was clear that no believable sea wall could keep this green sea out of the city.

Moreover, the Green Bar was now meters wide. Much could still be done to keep the sea at the bottom of this bar and not let it reach the top.

Eric hugged Yvette and said, "It's good."

The virtual crowd, at first frozen, broke into cheers and clapping. They then joined hands in long curves and jumped up and down generating ripples that spread across the Inner Harbor of Baltimore and gently rocked the boats.

At the Inner Harbor

It was now full dark and all the fancy lights of the Light City festival were on. People picked the glowing orbs from the trees and simply walked off with them.

A school marching band took over the street side walkway pushing through the crowd of people like a snowplow. The people had to stand on the planted boarder to let them pass; it was a bad night for plants overall. Behind the band flowed dozens of black teenagers dancing with the beat, ablaze with fairy lights; Eric knew that they are doing a much better job at dancing than he ever could, not even on this his very best day ever. But, Yvette could dance and did.

DIDACTIC ACTIVITIES

The suggested didactic activities focus on several variables as follows:

Key Theme

- *Technical measurement, control and forecasting of the climate change*

Pre-Teaching: It is a pre-reading moment useful for the introduction of ideas and key words present in the story, the verification of the vocabulary already known to the students and useful for understanding the text.

Reading: Descriptions or [if added] images play a key role. At first reading it can focus on the screenplay, postponing it to a second reading the analysis and understanding of the message. For the students, it certainly proves very engaging for them to actively participate in reading by miming some scenes. In this way, the understanding of meaning becomes more immediate and less memorization challenging.

Vocabulary and Several Issues Related to Technology/Science [Quotes as an Example]

- *"A fid is a tool for splicing rope, left over from the great days of sail. Baltimore has been married to the sea for a long time."*
- *"A fid is a hand spike," [...] "You force it through the tight strands of a rope so you can interweave the loose end strands through to make a splice. This art piece is a spiky tool too. A tool to help people comprehend climate change."*
- *"avatar"*
- *"terrorist"*
- *"shareholder"*
- *"laser"*
- *"barge-crane"*
- *"panic room"*
- *"cyberspace"*
- *"solar cells"*
- *"The green light bar and sea level continued to rise. By 2050 the water level was up the base of the Fid and was starting to cover the lower rows of solar cells"*

- *"The green bar got wider now because it also communicated the uncertainty of the calculation. Within that uncertainty was all that humanity could do."*

Presentation of the Characters: The exercise is impersonating characters [answering questions: What's your name, How old are you ? Where are you from? What are you wearing?], trying to act and interact as the character would have done. The dramatization proves to be a positive and productive activity. Moreover it encourages the student to express himself without inhibitions and allows him to customize a text:

Scenes: The teacher can ask the students to represent the different scenes of the story and write alongside them keywords. The story can be recited and recorded by creating a movie, writing an article for the school newspaper, and contributing to the creation of the school website.

Retelling the Story: Students attention is focused entirely on the meaning of story. If understood, they can be encouraged to change some actions or the ending of the story.

Skills Involved

- **Comprehension:** Understanding the story; listening to and identifying key words related to interactions between characters.
- **Interaction:** Participate in a recited version of a story through mime and repetition of words.
- **Image Education:** Graphically representing scenes from the story, using different techniques; creation of a comic book graphically depicting a personal reworking of the story.
- **Citizenship Education:** Collaborating within the class group, in a small group and in pairs; consolidating the identity of the class group; reflecting on scientific issues and the technical development of the society.

TECHNICAL ISSUES

Can Art Convince?

In this story, "The Fid", a major public piece of art is designed and selected by competition to clarify the problem of sea level rise to the public. The piece is both physical, a giant spike in the Baltimore Inner Harbor, and a virtual reality on line.

Should public art support a key idea? At what point does that just become propaganda? Can art convince? Can art generate buy-in for action on great problems?

Should We Expect a Silver Bullet?

Society faces a number of historic problems in the 21st century. Is it reasonable to expect that technology will provide a silver bullet that will solve them all?

A much more likely scenario (Powell, 2007) would be that a large number of different actions will be needed and that only a few will work very well and most turn out to be simply too little, too late. It will take an enormous amount of technical work to perfect those efforts that are worthwhile including the devoted efforts of a large number of people. Such efforts are called wedges.

The stories in the following two sections look at people who are working hard at two very different wedges, fertilizing the seas with minerals, and using cattle to drive back deserts. Just to make the stories more interesting, a high-tech concept, artificial Intelligence is included.

What other wedges are worth trying? Who will do the work? Will the reader be one of the people doing the work?

Figure 2. Baltimore Inner Harbor (photo by Riley)

CONCLUSION

The great problems of the 21st century, especially sea level rise, are best dealt with early. The longer society waits, the more drastic the corrective action will be needed.

Historically ecological movements have tried to frighten people into action. Fear does move people to action but only for a short time. Fear also tends to burn people out.

The far better approach is to generate buy-in. Buy-in is a mental state of human beings where the person experiences an idea from their leadership, envisions themselves succeeding with the idea, gets into action, and stays in action for a long time. Buy-in is the human mental state that built the Great Pyramids, built the Gothic Cathedrals, and launched Apollo to the Moon. Buy-in requires leadership, is by invitation only, and is primarily about vision.

A number of great 21st century problems can now be seen. Possible paths of action to address them can be envisioned. The invitations need to be issued.

This story suggests the use of art and video games to issue just such an invitation. Could this approach be effect? What other invitations to buy-in need to be issued?

Transition

This story tells of one odd and interesting way people might take on great social problems. It will take a great many people to take on the great problems of the 21st century. Can progress be made through art and games? Can technical people work with artistic people?

In the next story, "The Cabin Boy's Tale", a young person joints a team testing out one possible way to address global warming. Could some of the people who do this work be young people? Could one of these people be the reader? Is there a story in this effort? Could that story be the student's story?

REFERENCES

Berger, C., Meier, K. J. S., Kinkel, H., & Baumann, K.-H. (2014). Changes in calcification of coccoliths under stable atmospheric CO2. *Biogeosciences*, *11*(4), 929–944. doi:10.5194/bg-11-929-2014

Fabry, V. J. (2003). *Calcium carbonate production by coccolithophorid algae in long-term carbon dioxide sequestration*. San Marcos, CA: California State University.

Flynn K. (2016). The role of coccolithophore calcification in bioengineering their environment. *Proc. Biological Science, 283*(1833).

Joassin, P., Delille, B., Soetaert, K., Borges, A. V., Chou, L., Engel, A., ... Gregoire, M. (2008). A mathematical modelling of bloom of the coccolithophore Emiliania huxleyi in a mesocosm experiment. *Biogeosciences Discussions*, *5*(1), 787–840. doi:10.5194/bgd-5-787-2008

Meier, K. J. S., Beaufort, L., Heussner, S., & Ziveri, P. (2014). The role of ocean acidification in *Emiliania huxleyi* coccolith thinning in the Mediterranean Sea. *Biogeosciences*, *11*(10), 2857–2869. doi:10.5194/bg-11-2857-2014

Nanninga, H., & Tyrel, T. (1996). Importance of light for the formation of algal blooms by Emiliania huxleyi. *Marine Ecology Progress Series*, *136*, 195–203. doi:10.3354/meps136195

Powell, H. (2007). *Fertilizing the Oceans with Iron, Should we add iron to the sea to help reduce greenhouse gases in the air?* Retrieved from http://www.whoi.edu/oceanus/feature/fertilizing-the-ocean-with-iron

Sorrosa, J., Satoh, M., & Shiraiwa, Y. (2005). Low temperature stimulates cell enlargement and intracellular calcification of coccolithophorids. *Marine Biotechnology (New York, N.Y.)*, *7*(2), 128–133. doi:10.100710126-004-0478-1 PMID:15782289

Sorrosa, J., Satoh, M., & Shiraiwa, Y. (2005). Low temperature stimulates cell enlargement and intracellular calcification of coccolithophorids. *Marine Biotechnology (New York, N.Y.)*, *7*(2), 128–133. doi:10.100710126-004-0478-1 PMID:15782289

Young, J. R., Poulton, A. J., & Tyrrell, T. (2014). Morphology of *Emiliania huxleyi* coccoliths on the northwestern European shelf – is there an influence of carbonate chemistry? *Biogeosciences*, *11*(17), 4771–4782. doi:10.5194/bg-11-4771-2014

KEY TERMS AND DEFINITIONS

Buy-In: Buy-in is a mental state of human beings where the person experiences an idea from their leadership, envisions themselves succeeding with the idea, gets into action, and stays in action for a long time. Buy-in is the human mental state that built the Great Pyramids, built the Gothic Cathedrals, and launched Apollo to the Moon.

Climate Change: A term used instead of global warming. Indicates the change in the Earth's climate. This change is largely attributed to emissions in the Earth's atmosphere of increasing quantities of greenhouse gases (with a consequential increase in the greenhouse effect).

Fid: A tool used a device used by sailors to splice ropes. It is a metal spike with a wooden handle.

Virtual Reality: Entails presenting our senses with a computer-generated virtual environment that we can explore in some fashion. This could, of course, mean anything but it usually refers to a specific type of reality emulation.

Chapter 5
The Greenland Ice:
A Story of Changing Climate

ABSTRACT

Human society has done great damage to the oceans of Earth. It will take the dedicated efforts of a great many people working on a great many ideas to take on these great problems of the 21st century. Some of those people will be young people. In this chapter's story, "The Cabin Boy's Tale," a young person serves as an intern on a vessel involved in monitoring a mineralization of the seas program. In a few months, the young person learns more than he bargained for.

TOPICS FOR DISCUSSION

The following discussion points come from information covered in this chapter:

1. As the melting of Greenland's ice sheet continues, how will different societies around the world be effected?
2. What kinds of efforts might be tried to address the problems of the oceans? Of the melting ice caps?
3. What efforts will be needed to monitor and fine-tune whatever efforts human societies make?
4. Is there a story in place of young people in these efforts? Could that story be the reader's story?

DOI: 10.4018/978-1-5225-8401-8.ch005

Figure 1. S.S. Jeannette Expedition (1879-1881) in North Pole before encountering almost any ice, (Source: http://www.history.navy.mil/)

INTRODUCTION

The Oceans cover 70% of the Earth's surface, and hold 97% of Earth's water. And the glaciers currently cover around 16 million square kilometers of the surface of our planet and almost all are represented by the Antarctic ice sheet and Greenland. They are formed in places where it is very cold, more precisely where more snow falls in winter than melts in summer.

The sea ice is formed when sea water freezes excluding most of the salt at very low temperatures and to this is added also the snow. The sea ice that forms in Antarctica is seasonally melts back to land's edge each spring, while the core of the Arctic ice can last for several years.

In a glacier there are two zones:

1. The *accumulation area,* where the accumulation of ice occurs. In this area, the snow falling during the winter remains stored even in the hot season. In the high mountain glaciers, the accumulation zone is located at high altitude.
2. The *area of ablation,* where the loss of ice occurs. In this zone the melting of the fallen snow and ice occurs as well as detachment of parts of ice. Alpine glaciers have an area of just 3,000 square kilometers and make up just 0.018% of the Earth's surface; 46% are in Switzerland, 20% in Italy, 18% in Austria, 14% in France, and 0.03% in Germany.

Figure 2. Janette trip. (Source: http://nsidc.org/data/virtual_globes/)

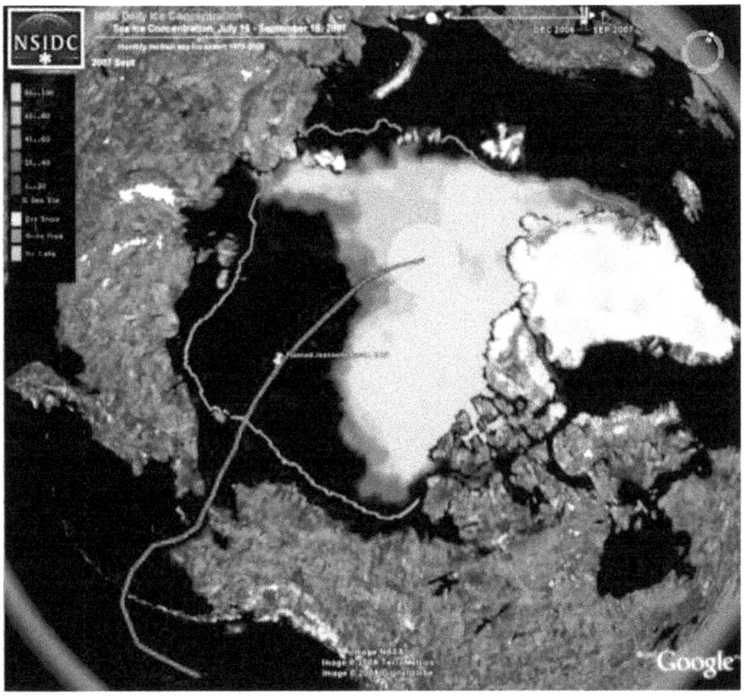

The largest glacier in the Alpine chain is located in the Bernese Alps, which has an area of 86 square kilometers and a length of 24 kilometers. About 50% of the glacial surface has disappeared in the Alps in 150 years.

Society can measure the melting of the Greenland ice sheet and monitor the polar ice caps. Society can measure the rise of global sea level to an accuracy of millimeters per year. Both of these are increasing at an accelerating rate.

According to NASA scientists, the Greenland ice sheet is melting faster than it is being replaced, contributing to sea level rise. From 1980 to 2000, the largest acceleration and mass loss came from southeast Greenland. From 2000 to 2007, the trend extended to include central east and west Greenland.

The main areas threatened by rising sea levels are coastal settlements, where regional changes can play an even stronger role than the global increase.

In 2005, 94% of Italian glaciers were falling back, 6% were stationary, none were progressing. At the Poles the situation is even more alarming.

In 2002, between January and March, an entire platform collapsed, an area of ice as extensive as the Valle d'Aosta disappeared in just 35 days, This involved an area of 3250 square kilometers with a thickness of 220 meters.

Figure 3. Sea Ice Loss (Source: https://www.erdc.usace.army.mil/Media/Images/)

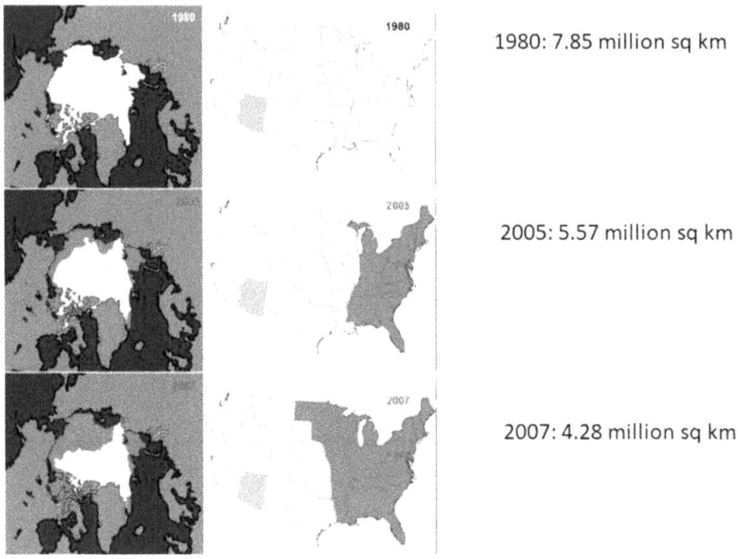

Figure 4. Ice Sheet Change 2002-2009 Derived From GRACE satellite observations (Source: Vasquez, J., 2015).

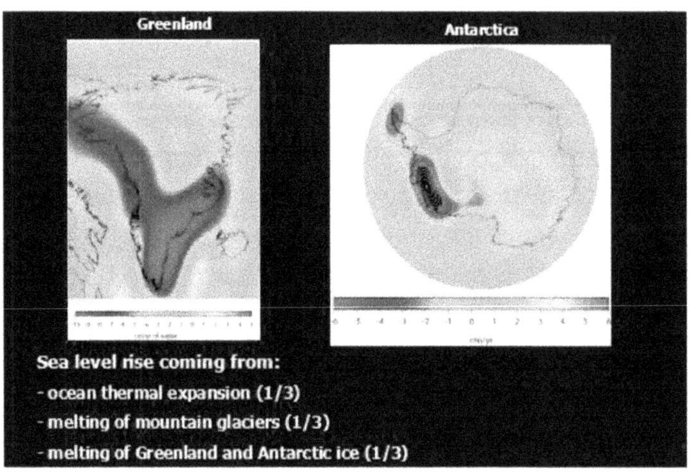

In 2012, on the other hand, the maximum reduction of the surface area of the Arctic summer ice pack was recorded since 1979, since when satellite surveys existed.

The Greenland ice sheet, the monitoring of the polar ice caps, the monitoring of the rise of global sea level to an accuracy of millimeters/year confirm that

Table 1. The more recent NASA projects

Name	Purpose
NASA's Ice,	Cloud and Land Elevation Satellite-2 [ICESat-2] will measure the elevation of ice sheets, glaciers, sea ice and more — all in unprecedented detail.
NASA's Making Earth System	Data Records for Use in Research Environments [MEaSUREs] data sets are intended to serve as Earth System Data Records, providing long-term broad-scale information about the cryosphere.
The NASA High Mountain Asia project	This project is using satellite observations, modeling, and field data to better understand changes in the region's ice, snow, and precipitation, and related human impacts.
Satellite Observations of Arctic Change	[SOAC] NASA satellite data and research on Arctic change, develops maps that illustrate the changes taking place in the Arctic over time.
The NASA Snow Ex campaigns	This mission is collecting multi-sensor airborne and in situ observations to develop and refine techniques for mapping terrestrial snow cover around the globe.

they are increasing at an accelerating rate (Vasquez, J., 2015). The satellites, moreover, confirm that the polar ice cap is shrinking more and more but the alarm cries of the Inuit people who live in these lands remain, unfortunately, almost always unheard.

The National Snow and Ice Center (NSIC, 2018) reported that in March 2015 the maximum extension of the winter frozen marine surface (which represents the period in which this extension widened and strengthened) reached the lowest value ever recorded since satellite recordings existed (14.280 million of square kilometers). Sea Level rises because of the addition of freshwater and thermal expansion Greenland ice sheet melting at an accelerating rate.

DIDACTIC ANALYSIS

The following story is focused on the trip through the Atlantic Ocean to monitor the effects of an ocean mineralization effort. It is analyzed didactically as fiction storytelling, highlighting how it can be used both to explore the interaction between Nature [Life], Artificial intelligence, and technologies. It addresses ethical issues and provides elements of problem solving management.

Environments

This story starts in the port of Annapolis, Maryland and then sets sail on an Atlantic cruse of the science ship, The Yvette A. Wight. One scene is also set in Kingston Harbor, Jamaica.

Main Characters

The main characters in this story are:

1. **Jace**: He is the cabin boy [CB] on the Yvette A. Wight. He is a summer intern and recent high school graduate.
2. **Yvette** or **The Yvette A. Wight**: This is both the boat and the artificial intelligence that sails it. She is at once a single-masted schooner of some 30 meters, an avatar in virtual space, and a legal person. She appears on monitors as a young African American woman.
3. **Dr. Carol Delanie**: She is the science officer [SO] on board, an oceanographer who works for Woods Hole.
4. **Captain Dodd:** He is the captain of the Yvette A. Wight and an officer in the merchant marine.
5. **Tuck**: He is an able-bodied seaman [ABS1], an ex-marine, and an excellent shot.
6. **Carter**: He is a waterman and fisherman, [ABS2] who knows the sea in his gut.

This is the crew of the Yvette A. Wight, a working ship that can broadcast mineral pellets into the ocean and take scientific measurements of their effect. The human characters are people supporting this effort despite personal hardship includes a young person just getting his sea legs.

Technology

The main cited technologies are: Virtual Reality, Artificial Intelligence applications, and mineralization of the oceans. For a specific in-depth discussion, see the dedicated Technology Section below which covers such questions as how humans and AIs could work together as a crew.

Communicative Approach

There is no doubt that in the communication process the oral language is of primary importance, as well as the starting point for learning a second language.

Stories motivate learning; they are fun and help develop a positive attitude especially towards the second language. The children are involved personally because they often identify with the characters and try to interpret the story and the illustrations thus favoring the development of creativity.

- Listening to stories helps children pay attention to rhythm, intonation, and pronunciation.
- Storytelling facilitates the development of learning strategies such as listening to grasp the general sense, predicting later scenes, guessing meanings, and hypothesizing.
- This story allows the teacher to introduce and revise words and structures, exposing the learners to the language and enriching their linguistic heritage, especially technical keywords.
- In the case of a foreign language, the narration offers opportunities to present cultural information of the country in which the language is spoken. Every word has its own history, its particular meaning, its value that is not always easily translated into another language but that with the right material we can try to "pass" to the students, immersing them in the atmosphere of the second language.

Figure 5. Model of the Yvette A. Wight (personal image by Riley)

Playful Approach

It is therefore important to cut out moments dedicated to activities within the class substantive games and activities, capable of producing true emotional involvement, a fundamental engine of any form of learning.

Transition Into the Story

Human society has done great damage to the oceans of Earth. Are there things that can be done to address these problems?

It will take a great many people to take on the great problems of the 21st century, like the state of the seas and air. Could some of those people be valent young people? Could one of these people be the student? Is there a story in their effort? Could that story be the student's story?

THE FICTION STORY: "THE CABIN BOY'S TALE" BY TOM RILEY

Premise

Sometimes you simply must go to sea, particularly if the sea is coming to you.

Annapolis Arrival

Coming to a stop, the bus made that irritating whine that electric drives do when they have run up too many kilometers. Jace got off the bus and with a few grunts and pointing, retrieved his duffle bag then took his phone out of his purple and orange vest. The Nav App showed he was in downtown Annapolis and only a couple of kilometers from the Iron Seas wharf. It was good to walk after sitting for so long. The early June air was warm and the sun felt good.

He passed a pair of midshipmen in their too-crisp uniforms and nodded. He had decided against applying to the military academies; they were still fighting last century's wars.

He thought about getting some lunch but all the cafes he passed were stuffed with old tourists. He hated tourist traps and he hated the company of old people even more. He settled on an energy bar from his vest pocket and a large ice cream from a shop that opened on the street.

When he got down to the dock area he could see that several blocks of ancient buildings in the low areas had flooded one too many times and were now being demolished.

The Yvette A. Wight was easy to spot. She was larger than most of the sailing ships in harbor and had a tall, proud, black wing for a mainsail. It also helped that she had the words “Iron Seas” boldly displayed across her dark wing.

Jace had to squeeze past a rental truck that was unloading heavy sacks. He nodded at the two men unloading the bags with large single letters, “A”, “B”, “C”, written on them. Giving a nod, he continued walking down toward the boat.

At the boat’s rail stood a man in work clothes but wearing a white weathered captain’s cap and holding an e-tablet. He was using it to check in the material from the truck. Jace recognized Captain Dodd from the godparent events they both had participated in over the Web but he had never actually met the man in person.

“Ahoy there,” said Captain Dodd, “You would be our new CB. I’d know you anywhere.”

“Yes,” replied Jace.

“Yvette spotted you five minutes ago,” continued the Captain, “Never underestimate what she can see with her High Eye.” The captain gestured at the instrument pod at the top of the mainmast.

He held up his e-tablet that now showed an animation of a young African American woman with a feathered cap. She was Yvette, the ship’s Artificial Intelligence. Jace had interacted with her many times in the hundreds of hours of virtual training he had completed in the last year.

“First thing, you need your ID,” continued the captain while taking a lanyard from his pocket and tossing it to Jace.

“Wear this whenever you are on board,” he continued, “The fob has a transponder so Yvette can track you whenever you are within a few klicks of the boat. It also has enough memory in it for your personal log of the cruise.”

Jace put the lanyard around his neck and slipped the ID device hanging from it into an upper vest pocket. He always wore some electronics but not as many as some of his contemporaries. However, one device more or less did not matter much. Yvette then rang his phone and directed him to the forward hatchway.

"We will have the formal introductions at the dinner on shore this evening," said Yvette, "Until then stow your gear and then help the best you can with the loading. We sail with the tide at 0400 hours in the morning."

On this Iron Seas expedition, the crew consisted of five people and an artificial Intelligence [AI]: Captain Dodd; Dr. Carol Delanie, the science officer [SO, from Woods Hole]; Tuck [an able-bodied seaman and ex-marine, ABS1]; Carter [a waterman and fisherman, ABS2]; and, of course, Jace, the cabin boy [usually called CB]. The Yvette A. Wight was both the boat and the artificial intelligence that sailed it. She was at once a single-masted schooner of some 30 meters, an avatar in virtual space, and a legal person.

First Day at Sea

Jace's first day dawned bright with gently rolling seas. The sea air was fresh and the wind out of the north. The entire crew gathered within earshot of the small galley for the morning briefing.

"Good morning," Yvette said, "The weather is clear and we are approaching the edge of our next chum area. Here is our broadcasting plan." On the main monitor, she stood in her iridescent feather cap beside satellite images of the North Atlantic. She then zoomed in the image to show their position and course, with the chum area blocked out in blue.

"Today, we will be chumming four lines," picked up the science officer. "Each will start with one hour and 15 minutes of 'A' and then finish with 45 minutes of 'F'. We will be chumming by hand."

"As this is Jace's first line," said the Captain, "let me go over the process. First 'A' is our standard formula for fertilizing the microbes that take carbon from the water and make chalk. Think of the White Cliffs of Dover. Do it wrong and you could get red tides and about a billion jellies."

"The two Able Bodied Seamen and the CB will act as a team to bring up the sacks of pellets from the hole and then broadcast the material over the transit one scoop at a time. You three should rotate jobs so Jace gets to try every position. I am counting on the ABSs to show Jace the ropes."

"The 'F' material is a different kettle of fish, so I will let the Science Officer tell you about that one."

"This is a normal distribution on a beautiful day," the SO took over, "so don't screw it up. The 'A' will feed the micro-organisms and will tend to stay suspended in the surface waters."

"The 'F' is a special blend and quite experimental. We will be passing over a field of methane frozen in ice in the ocean floor, called methane-hydrate. We want very much to keep it there. The 'F' mix is designed to help stabilize the bottom microenvironment that keeps the methane frozen. It is heavier than the 'A' and it will sink like a stone, or rather, like many small stones."

"All done?" asked the Captain, "Let's get to it."

Carol, the SO, was hot, certainly the exact opposite of Dr. Spock. She was at least ten years Jace's senior but he did not care. There had been no indication that she was the captain's girl, or at least none he saw, and he was up for the challenge. She worked for Woods Hole Oceanographic Institution, not Iron Seas; she had a PHD in oceanography, and she always wore crummy work clothes. The only thing that bothered him was her drab clothes. The entire operation, including Yvette's ticket to operate at all, hinged on her science being fully trusted by the bureaucrats back on shore.

"Follow me," said Tuck, the ABS1, to Jace. He turned to the monitor and asked Yvette for the first sack. She displayed a layout of her holds with the chosen sack shown in red.

"If you don't get the right sack," Tuck continued to Jace, "The ship could get out of balance. Yvette does not like that and will be fast to tell you if you screw up."

They worked their way to the forward hold and found the sack. Yvette was shining a light on it.

"Take care," said Tuck, "it's heavy, twenty kilos."

Jace managed to get it on deck and to the stern cockpit even if he staggered a little doing it.

"Good," said Tuck, "but turn it so Yvette can see the letter." He pointed toward the monitor on the back of the cabin just to starboard of the aft hatchway.

Carter, the ABS2, had retrieved two aluminum scoops and then cut open the top of the sack. He showed Jace how to tie the line from the scoop handle to his wrist. They then waited for the countdown clock on the monitor to sound a bell.

Bong. Carter shoved his scoop into the sack, drew out one measure, and flung it over the transit, making a wide arc of tiny splashes on the gently rolling sea. He then waited for the bell again and gestured to Jace. Every two minutes, a wide arc of tiny splashes erupted.

After thirty minutes, they rotated jobs: port caster, starboard caster, and fetcher. On his second rotation as fetcher, Jace ducked into the cabin to bring up water for his new team. He slipped by the captain at the wheel to the port of the aft hatchway and stopped for a minute to talk to the SO, who was in her miniscule lab.

"Hello," Jace said while pulling himself up to his full height. "How are we doing on the line?"

"I'm not monitoring your work," Carol said. "You will have to speak to Yvette for that. I'm planning the science a couple days out."

"OK," Jace said.

"The real science is done a few weeks after the lines are broadcast," she continued. "That gives the minerals time to take effect on the phytoplankton."

"In two days, we will cut the wake of the Rachel L. Carson from a month ago. She is a much bigger ship than we are and can lay pellets by the ton. That said, she is really too big and expensive to operate, to be used to work down close to the water and take the detailed science we can. Also, she cannot really do the small jobs like the 'F' lines we are doing today."

The monitor first showed the coastal freighter Racheal L. Carson and then her iron lines, which animated to show their drift since they were broadcast.

"I'd better let you get to it," said Jace and took up a water jug.

First Night at Sea

That night was calm and for the first time Jace heard the whales sing. The distant, haunting sound simply came through the boat's hull. It was the loneliest sound he had ever heard. The whole boat was quiet, listening.

Second Day of the Third Week at Sea

The second day of Jace's third week dawned with a truly ugly sky and the seas were up. A gusty wind blew from the northeast. Yvette spent the whole morning briefing on the weather. She and the captain decided to stop all chumming, and science, and to run before the wind out of the storm's path. They set a new course due south.

"Jace, your life vest is in the upper rack on the port side," said the Captain, "If you see Yvette wearing one, put it on."

"Set the storm foresails and stow the Zodiac," ordered the Captain.

By 0900 hours, Jace was feeling cooped up in the stuffy cabin. He put on his life vest and exited aft. The captain was at the wheel.

"How you doing?" asked the Captain.

"Just fine," replied Jace.

At that moment, he suddenly realized that he would feel a whole lot better if he cleared his stomach. Jace made it to the rail.

"I guess I'm not such a great seaman," conceded Jace, looking and feeling rather green.

"Happens to everybody," said the Captain. "What counts is being able to eat tomorrow. Don't eat today. Eat a little tomorrow but only bland food. For now, sit down and keep your eye on the horizon."

"It helps if you hold your nose when you eat," volunteered Tuck.

Once the Yvette A. Wight had been a graceful ship, if under a different name. Then she could run with the best of them, a rich man's toy. Now with her add-on science modules and several tons of iron pellets in her holds, she labored in the swells and made no more than half her old top speed.

Jace tried to keep his eyes on the horizon but the rolling gray-green sea kept getting in the way. In between bouts at the rail, he was struck by the absolute power of the sea, how unbelievable it was that people could do anything that could harm it and the absolute absurdity that their small boat could do anything to help it.

After a long night, Jace woke to a steady boat on a rolling sea. He was able to keep down a piece of dry toast and to stand straight at the rail facing the sea. For the first time he understood that the sea needed him and that there would be personal costs, but costs that he would gladly pay. With consuming that simple piece of toast, he had assured his career on the sea.

The Yvette A. Wight headed into Kingston, Jamaica, for repair parts and supplies. For a while, Jace watched a flock of brown pelicans flying in a rough "V". They were big heavy birds who clearly knew where they were going.

Kingston Harbor

They entered Kingston Harbor Jamaica in the late afternoon. Everything still looked wet, and green tree foliage lay about. Fortunately, they found that the storm damage to the town was light. The captain soon appeared in his full dress merchant marine uniform, his dress blues. He announced that the supplies would not be available until the next day and everyone could spend the night on shore if they liked. They all, except Yvette of course, piled into the Zodiac. From the dock, the captain left to pay a courtesy call on the harbormaster; all the other crewmembers dispersed through the wet town.

Jace walked off by himself. For a long time, he saw only tourist traps and just like Annapolis they were full of old people. Eventually, he found his way down to a sailor's bar.

As soon as Jace was inside, he liked it. Strings of lanterns with dim LED bulbs hung throughout the rustic lounge. Stools stood along the bar that was near the door. An attached room on the far side of the building that was separated by a hanging fishnet housed pool tables. A semicircle of couches and lounge chairs with small tables between them for drinks filled the main room.

The semicircle surrounded a dance floor. Jace didn't find it appealing to join in with the old folks shimmying around to songs almost three decades old, but here he could finally sit back and sip on his drink.

Despite the sip he almost spat the drink back out. He wasn't expecting the bite of the booze.

"Shots are better to start with, love." A woman about his mother's age stood just to his left as she balanced a tray with a collection of drinks of all colors and sizes. She was wearing a short red dress with a frilly skirt and black lace along the hem. The top part of the dress was so tight her boobs looked about to pop out. Blonde curls framed the older face, but time had been kinder to this woman than most.

The woman smiled and bent down, offering him a better view of her chest as she handed him a shot glass. "Just chug it down in one go, hon."

He smiled back at her as he took the drink and lifted the glass to his lips. They didn't break eye contact as he tipped it back and gulped it all in one swallow. His throat felt like it was on fire but he wasn't about to let the woman, whose nametag said Brandy, see him flinch.

She was okay for an old broad and maybe if he played his cards right ...

Then there stood Captain Dodd, big as hell in a dark blue uniform.

The shot glass dropped from Jace's hand, making a loud bang on the table. Brandy's gaze wasn't on him now, but on the big man standing in front of him.

She wasn't smiling anymore.

"Servin' a minor." It was the Captain's gruff voice. Jace couldn't look up; that would mean looking his godfather in the face, but he could imagine the steely look in Dodd's ice-blue eyes.

"Ye wouldn't be trying to take advantage, would ye?" As the Captain spoke, Jace felt the rush of warmth in his face and glanced up at Brandy, who backed away and muttered an apology for any misunderstanding.

"Finish your drink, son," said the Captain, "Say something polite to the lady, and pay your bill. Don't skimp on the tip either."

Neither said much more about the incident. The whole crew worked hard the next day.

Back at Sea

In Jace's fifth week, on a bright summer's day under sail with a light breeze, Yvette spotted a cigarette boat in the distance. Fast and powered with a huge gasoline engine, these boats were favored by smugglers.

Once, these boats had been a threat to Iron Seas. Early on, there was an incident where smugglers took over a boat, set the crew adrift in the Zodiac, and used the boat to run several tons of contraband before abandoning it.

Now Yvette spotted such boats with her High Eye and automatically sent full documentation to the Coast Guard. The smugglers knew the new drill, so the cigarette boat was soon out of sight. The Yvette sailed on.

"I think we can train Jace to repel boarders today," said the Captain, "I was not too impressed with his early sims."

Tuck fetched several empty plastic jugs and a length of twine. He tied a jug on the line at ten-meter intervals and trailed the line over the transit.

Carter took the helm and the captain disappeared below. He reappeared with an assault rifle and two nine-millimeter handguns. He handed the rifle to Tuck.

"First let's review Jace's early training for the edification of the entire crew," said the Captain, "Yvette, run the tape."

Yvette's monitor read "Searching." The SO adjourned to her repel-boarders station at the main computer terminal in the cabin. She was glad to be out of the way. Then every monitor on the ship started playing Jace's early training file from some months before.

On a Virtual Sea

"Pirates. Just off the port bow."

"How far out?" The captain did not reach for the binoculars that Jace extended. He pulled something out from his waistband and when Captain Dodd stood up Jace saw it was a gun.

"Can't we outrun them?"

"They're too close."

Jace scrambled to remember the training. What had they been taught to do if they saw pirates? Where was everyone else anyway?

"Do you want me to do an all call?"

"Everyone who's left is here."

"What?"

"You saw the SO swept overboard. The Tuck pulled you in but struck his head on his way below deck. He died."

"And Carter?"

The captain nodded toward the bow and Jace turned to see the scraggly hand already armed and on the deck. "Stay here."

Captain Dodd yanked the door open and didn't worry about closing it behind him. The pop pop of the rifles was followed by a shriek and Jace watched as Carter fell. Captain Dodd was taking aim and Jace tossed the binoculars on the bench and scrambled out to aid Carter.

He didn't need to take a pulse. Carter's blank eyes stared out at nothing and his fingers still gripped his gun. Jace pried them off and started to stand up, but the captain turned and knocked the weapon out of his hand.

"They won't shoot you."

"But they will shoot you. What am I supposed to do then? Swim?"

Captain Dodd's steely blue eyes had a wistful look. "Don't you get it? To them, you're worth as much as the boat."

A clap rang out and Dodd's chest exploded as red blood sprayed across his white shirt. Dodd fell to the deck and Jace sank down beside him.

What was he going to do now? He'd blown off most of the training and couldn't steer the boat to save his life. Even if Yvette could help him navigate, he probably would not make it home. Moreover, none of that mattered because there was no way the pirates would leave him in peace on an intact ship. Not one of these science vessels. It would be stripped for parts to be sold on the black market while he was held for ransom.

"I failed." He looked down at Dodd's eyes. Dodd groaned and reached up with a hand that Jace clasped. Jace took a deep breath. He couldn't fail. Pirates or no pirates, come hell or high water, he had to try.

A raspy long rattle escaped Dodd's mouth as he died and Jace looked up to see a man dressed head to toe in black standing over him. The man raised the butt of the rifle and struck Jace in the head with it.

On a Real Sea

Jace was mortified.

"Clearly we have here a well thought through example of how to handle boarders," said the captain. He then pulled the handgun from his waistband and slipped in a magazine. He clicked the safety and using a two-hand grip fired three shots. The first two shots bracketed the second jug, the third shot plugged it.

The captain then showed Jace how safely to handle a gun on a boat. Jace managed to hit one jug by expending the rest of the nine-round clip.

The captain nodded to Tuck who shouldered the rifle and picked off all the remaining intact jugs one after the other, working from close to far. Yvette logged all the slugs as metal thrown into the sea.

Work at Sea

Later that week, the SO scheduled a trawl. The biggest net they could handle was not very big but it was big enough to take a few marine samples.

"Go'n fishing!" The Captain liked the idea.

Jace's team deployed the sampling net from the port science compartment and paid it out. Yvette fired up the engine to add the motor's power to the wing.

They towed the net for an hour. Jace's team then hauled it in and emptied the contents into the transit well.

The SO sorted, identified, and logged the specimens. The captain brought out his own bucket and the CO tossed an occasional specimen into it, mostly dead squid. The rest went back into the sea. A flock of raucous, hungry gulls had somehow found them.

After three trawls, the SO called the test complete. Jace's team then cleaned the equipment.

"All done?" The captain appeared from the cabin carrying a deep-sea rod and reel. He took a squid from his bait bucket and hooked it.

For the rest of a long afternoon, the captain fished. Carter gaffed the fish and pulled them over the transit. The SO weighed and logged each fish caught.

Jace held the rod for a while but didn't get a strike, but at least he didn't drop the reel into the sea.

They ate well that night.

Headed Home

It was late summer and hot when they dropped Jace off in the Florida Keys. Part of his home trip would be by ferry but mostly it was just a long bus ride. The ride gave Jace time to think.

Jace now knew he was a seaman at heart: The seas have rolled for a very long time, but somehow we had managed to damage them in less than 200 years. The seas were here long before us and they will be here long after us. The least we can do is clean up our mess and thereby let the seas roll on. If that hard job had fallen to him, so be it.

DIDACTIC ACTIVITIES

The suggested didactic activities focused on several variables as follows:

Key Theme

- *Protection of our seas. Support of artificial intelligence in technical activities at sea.*

Pre-Teaching: It is a pre-reading moment useful for the introduction of ideas and key words present in the story, the verification of the vocabulary already known to the students and useful for understanding the text.

Reading: Descriptions or [if added] images play a key role. At first reading it can focus on the screenplay, postponing it to a second reading the analysis and understanding of the message. For the students, it certainly proves very engaging for them to actively participate in reading by miming some scenes. In this way, the understanding of meaning becomes more immediate and less memorization challenging.

Vocabulary and Several Issues Related to Technology/Science [Quotes as an Example]

- *The crew consisted of five people and an artificial Intelligence [AI]: a Captain, the science officer [SO, from Woods Hole]; an able-bodied seaman and ex-marine, ABS1, a waterman and fisherman, ABS2, the cabin boy.*
- *The Yvette A. Wight was both the boat and the artificial intelligence that sailed it. She was at once a single-masted schooner of some 30 meters, an avatar in virtual space, and a legal person.*
- *A small galley*
- *Satellite images of the North Atlantic*
- *our standard formula for fertilizing the microbes that take carbon from the water and make chalk*
- *The White Cliffs of Dove.*
- *red tides and about a billion jellies*
- *a field of methane frozen in ice in the ocean floor, called methane-hydrate*

Presentation of the Characters: The exercise is impersonating characters [answering questions: What's your name, How old are you ? Where are you from? What are you wearing?], trying to act and interact as the character would have done. The dramatization proves to be a positive and productive activity. Moreover it encourages the student to express himself without inhibitions and allows him to customize a text:

Scenes: The teacher can ask the students to represent the different scenes of the story and write alongside them keywords. The story can be recited and recorded by creating a movie, writing an article for the school newspaper and contributing to the creation of the school website.

Retelling the Story: Students attention is focused entirely on the meaning of story. If understood, they can be encouraged to change some actions or the finally of the story.

Skills Involved

- **Comprehension:** Understanding the story; listening to and identifying key words related to interactions between characters.
- **Interaction:** Participate in a recited version of a story through mime and repetition of words.
- **Image Education:** Graphically representing scenes from the story, using different techniques; creation of a comic book graphically depicting a personal reworking of the story.
- **Citizenship Education:** Collaborating within the class group, in a small group and in pairs; consolidating the identity of the class group; reflecting on scientific issues and the technical development of the society.

TECHNICAL DISCUSSION

Is Personal Experience a Key?

In the Iron Seas stories, the small but critical scientific expeditions go to some trouble to include a member of the public on board. In one story, it is a cabin boy and in another, it is a guest cook. The reason implied is to build up a cadre of people who have personal experience with the sea fertilization effort and to promote openness.

How important are personal experiences in understanding and supporting a life's effort? Would the student go on such a cruse?

Figure 6. Plankton bloom from Space [NASA] (Source: NASA, 2018)

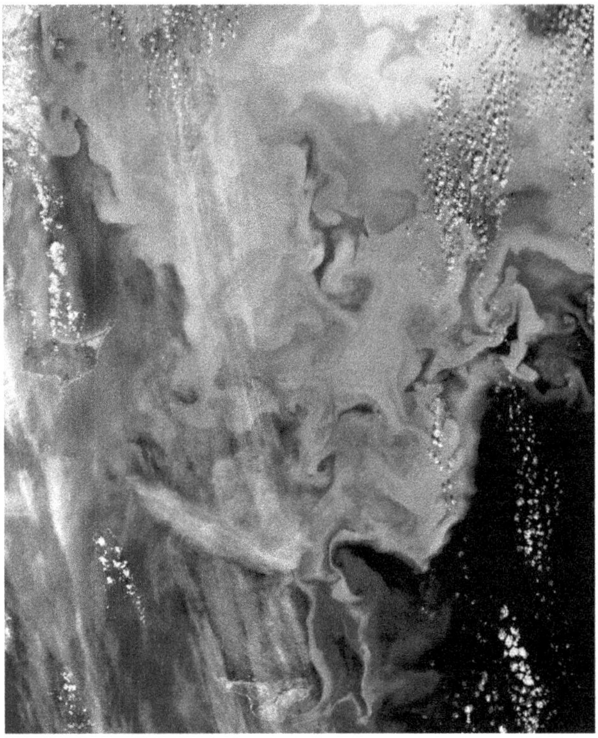

What minute living things in the sea might help remove carbon dioxide from the air?

Phytoplankton: Minute plants that live in the ocean. Many have skeletons made of calcium carbonate, notably the Coccolith. After the cells die, these solids can settle to the bottom thus sequestering a very large amount of carbon. This opens up the possibility of making a major contribution to addressing global warming. The plants grow seasonally over wide expanses of the ocean and their seasonal growth is visible from space. Their growth is naturally limited by the availability of minerals in the water, particularly iron.

It may be possible for people to fertilize the minute plants of the oceans with a few minerals to greatly increase the amount of carbon dioxide tied up. This effort would require a lot of very detailed information about the ocean environment if it were to succeed.

If the skeletons settle into very deep water, the high pressure there will force the calcium carbonate back into solution. The carbon sequestering will then only be temporary.

Adding minerals to the ocean to greatly expand their growth has been tried, but only on a modest scale. The first project organizer was not able to obtain the proper permits. This test was then conducted in secret and not enough scientific data has since become available to support larger scale tests.

A onetime large test was later run on the open ocean very far from land. The resulting plankton bloom was visible from space for about three weeks. This test was over very deep waters where the sequestering of carbon was at best temporary.

The phytoplankton are food to small animals that are intern food for fish. This opens the prospect of paying for the operation with increased fish harvest. Currently there is no legal structure to support the economics of such an open-ocean fishing operation. Ships from many nations can harvest the bounty of fish without paying for the mineralization project. This farther leads to secrecy.

Any attempt to sequester carbon in the oceans with mineralization will require much more testing and continuous monitoring from the surface, on the bottom, and from space. This process will exacerbate many political and international considerations (Adilee Mitra, 2018).

CONCLUSION AND FUTURE RESEARCH DIRECTIONS

Human beings have done quite a bit of damage to the oceans. Correcting that damage will not be easy nor can the effort be fully automated. People have disrupted a critical balance and it will take real work to put it right again. The extent of the problem has now been detailed in four major reports (IPBES, 2019; IPCC, 2019; Laybourn-Langton et al., 2019; NCA4, 2018).

Much of that work is scientific. The big data from satellites is critical but not enough in itself. First the big data must be analyzed, which requires significant computer power. From this analysis specific hypotheses on what is wrong and what can be done about it must be developed.

However, remote sensing data is not enough. A substantial effort is needed to put people and test equipment on and beneath the water. Both AIs and real people will be needed on site for this hard work.

Corrective ideas will then have to be tested on a large scale. The oceans are very big. There is much work to be done and much of that work is very controversial.

Transition

Whatever is done to fix the sea and air, the effort must be watched very carefully. This effort will take the dedicated efforts of a great many people. Some of those people will be young people.

In the next chapter, "The Captain's Tale", sometimes things will go wrong. If a great many people are watching just what is being done then, when a problem occurs, a great many people will want to argue about what to do next? Is there a story in the conflict when things go wrong?

One more story in this series can be found at the Big Moon Dig Web site (Riley, 2017).

REFERENCES

IPBES. (2019). *Intergovernmental Science-Policy Platform on Biodiversity and Ecosystem Services "Nature's Dangerous Decline 'Unprecedented', Species Extinction Rates 'Accelerating.'"* Retrieved from: https://www.ipbes.net/news/Media-Release-Global-Assessment

IPCC SR1. (2019). *The Intergovernmental Panel on Climate Crisis, Special Report 1.5.* Retrieved from: https://www.ipcc.ch/sr15/

Laybourn-Langton, L., Rankin, L., & Baxter, D. (2019). *This is a Crisis, Facing up to the age of environmental breakdown.* Retrieved from: https://www.ippr.org/files/2019-02/this-is-a-crisis-february2019.pdf

Mitra, A. (2018, April). The Perfect Beast. *Scientific American, 318*(4), 26–33. doi:10.1038cientificamerican0418-26 PMID:29557947

NASA. (2018). *Plankton Bloom from space.* NASA. Retrieved from: https://www.bing.com/images/search?view=detailV2&ccid=lXP5IMwN&id=FEB9F7CA7D0979D7714E63D18C39AF5989E0D1BA&thid=OIP.lXP5IMwNaab-yGwDmLhuowHaI4&mediaurl=http%3a%2f%2fscitechdaily.com%2fimages%2fNASA-Ocean-Data-Shows-Movement-of-Plankton.jpg&exph=2304&expw=1920&q=nasa+plankton+bloom&simid=608029271280124865&selectedIndex=18&ajaxhist=0

NCA4 Fourth National Climate Assessment. (2018). Retrieved from: https://www.globalchange.gov/nca4

NSIC National Snow and Ice Center. (2018). Retrieved from https://nsidc.org/

Riley, T. (2017). *The Big Moon Dig*. Retrieved from http://bigmoondig.com/BigMoonDig.html

Vasquez, J.A. (2015). STEM--Beyond the Acronym. *Educational Leadership*, *72*(4), 10-15.

KEY TERMS AND DEFINITIONS

Artificial Intelligence (AI): The intelligence shown by machines or computer software. An AI in some ways can mimic human intelligence but does not have to match it feature for feature. The interface avatars of these AIs are important characters in these stories.

Big Moon Dig (BMD): A fictional grassroots space movement with the purpose of building a settlement on the Moon. It is a MOVE organization. Historically Apollo to the Moon provided an enormous number of people with a vivid, positive vision of the future. Such a vision would be a major asset in addressing the great problems of the 21st century if it can be reestablished. The Big Moon Dig's top-level purpose then is to rebuild that vision.

Iron Seas: A fictional organization that is testing the seeding the oceans with iron to sequester carbon.

Chapter 6
Ocean Acidification:
A Problem for Policymakers

ABSTRACT

Absorption of CO_2 by the waters of the oceans is making them acid, which is then detrimental to much ocean life like corals. There are technological solutions to the problems of the oceans; action is possible. Nevertheless, whatever is done it must be watched very carefully. This monitoring effort will take big data taken from space and from the air, but it will also take a major hands-on effort by people right at the water's surface. This chapter's story, "The Captain's Tale," looks at what might happen when an effort that is closely watched by a great number of people runs into problems.

TOPICS FOR DISCUSSION

The following discussion points come from information covered in this chapter:

1. The coral reefs of the Earth's oceans are suffering from global warming and acidification of the water by carbon dioxide. Who does this hurt? What can be done?
2. There is always the possibility that things will not go right. Is there a story in this effort to get things right? Is there a story in the conflict that occurs when things go wrong?
3. It takes a village to raise a child, but what does this mean in a digital age?

DOI: 10.4018/978-1-5225-8401-8.ch006

INTRODUCTION

Water acidification is mainly caused by CO2 emissions which, when dissolved in water, generates carbonic acid. It is a process very similar to what occurs for the production of carbonated mineral water. The result is that the chemical composition of water changes, with negative consequences for the whole marine ecosystem.

The increase in carbon dioxide in the atmosphere, besides being one of the main causes of the increase in global temperatures, is also affecting the equilibrium of marine ecosystems. In fact, the oceans absorb between a third and a quarter of all the CO2 released each year in the atmosphere. While this phenomenon contributes to slowing down the negative consequences of the greenhouse effect, on the other hand it makes the waters increasingly acidic.

The negative consequences of water acidification on the marine ecosystem are manifold: coral reefs at risk, reduction of fishing resources and marine habitat compromised are some of the results that the change in the chemical composition of the seas involves.

Coral reefs are submarine forests. Moreover, forests that are rich in living species. Consider that at least 25% of all marine species in the world live in the waters of coral reefs. Moreover, what most upsets and attracts tourists is the magnificent combination of plants and fish that can be found in these places.

The majority of the coral reefs in our ranking - and in the world - are actually made up of many smaller elements, connected in a single ecosystem. The most important coral reefs in the world are:

- **Great Barrier Reef Australia**: This ecosystem reaches 2300 km in length and includes thousands of coral reefs and hundreds of islands with over 600 types of hard and soft corals. The barrier is home to countless species of colorful fish, mollusks and sea stars, as well as turtles, dolphins and sharks.
- **Red Sea Coral Reef**: These were formed by vast coral fields, which in turn are made up of tiny coral polyp colonies.
- **Coral Reef of New Caledonia:** This is a World Heritage Site by UNESCO since 2008 as "Lagoons of New Caledonia". It boasts more than 9,300 marine species.
- **Mesoamerican Reef, Yucatan, Belize, Guatemala and the Islands Of Honduras Bay:** A World Heritage Site by UNESCO since 1996, it

is the largest coral reef in the Atlantic Ocean. This coral reef extends almost 1,126 Km. Over 500 species of fish and 65 types of coral live within this great barrier system. Coral reef fisheries include snappers, groupers, and lobsters.

- **Coral Reef of The Maldives Islands, Indian Ocean:** Besides the corals, there are a large quantity of fish species: anemones, crustaceans, mollusks, clams and octopuses. The echinoderms are also very widespread.
- **Coral Reef of Apo, Philippines:** It is the second longest continuous coral system in the world and the largest in the country. The barrier and the waters that surround it are protected areas in the area administered as Apo's coral reef nature park.
- **Belize Barrier Reef, Caribbean Sea:** This coral reef is protected by the UNESCO program. It is the largest coral barrier in the northern hemisphere and second only to the great Australian coral reef. Also part of the natural system of the Belize Barrier Reef are atolls, beaches, lagoons, and mangrove forests. There are endangered species such as manatees, sea turtles, and American crocodiles.
- **Saya De Malha, Indian Ocean:** This marine habitat facilitates the life of particular species such as turtle and blue whales.

Figure 1. Coral Reef of New Caledonia (Source: https://www.forbes.com/sites/priyashukla/2018/08/17/new-caledonia-is-taking-new-steps-to-protect-their-coral-reefs/)

- **Coral Reef of Andros, Bahamas:** It is famous for its deep-water sponges and large schools of red snapper. It is also the only place in the world with more than one Nassau grouper aggregation, marlins, sailfish, stingrays, reef sharks, lobsters, turtles, moray eels, octopi, and game fish.
- **Florida Keys, United States:** It is the third largest coral reef in the world with about 1,400 species of marine and animal plants, including more than 40 species of rocky corals and 500 species of fish.

An estimated 30–40% of the carbon dioxide released by humans into the atmosphere dissolves into oceans, rivers, and lakes. The process is:

- When dissolved in water, carbon dioxide forms carbonic acid that primarily dissociates into bicarbonate and hydrogen ions.
- Some of the excess hydrogen ions combine with carbonate, decreasing carbonate availability to marine organisms.
- Carbonate is used by marine organisms like the pteropod [a marine snail] to create the compound calcium carbonate.
- Calcium carbonate is the material that composes the shells and exoskeletons of many marine organisms.

Too much carbon dioxide in the ocean has the potential to harm marine organisms and ecosystems. As the concentrations of CO2 dissolved in the water increase, there is a more vigorous growth of the plant life forms, the same phenomenon that also occurs on the land. On the contrary, animal life requires large quantities of oxygen even under water, an element that pushes the fish and other life forms to move away from the ocean layers with more vegetation. Therefore, human activity is creating layers, above all at greater depths, without animal life because the conditions are insufficient to carry it forward.

Human activities release carbon dioxide into the atmosphere. Burning fossil fuels adds extra CO2 to the atmosphere. 24,000,000 millions of CO2 lbs was released from fossil fuels in 2011.

Dissolved CO2 reacts to form carbonic acid, lowering the ocean's pH. As more CO2 is added to the air, more dissolves into the ocean and more converts to H2CO3.

To aggravate the situation there are two further serious phenomena. Seas and oceans play a fundamental role in keeping the temperature on the mainland stable, absorbing about 90% of the planet's heat and acting as a climate regulator thermostat. Because of the climate changes that are causing the warming of the atmosphere and ocean waters, this delicate balance could be compromised. Furthermore, the raising of the level of the same waters, is being set in motion by the melting of the glaciers.

DIDACTIC ANALYSIS OF "THE CAPTAIN'S TALE"

The following story is focused on a trip over the Atlantic Ocean. It is analyzed didactically as fiction storytelling, highlighting how it can be used both to explore the interaction between Nature [Life], Artificial intelligence, and technologies. It addresses ethical issues and provides elements of problem solving management.

Environments

Nearly this entire story takes place on the scientific research vessel the Yvete A. Wight on a cruise in the Atlantic Ocean.

Main Characters

The main characters in this story are:

1. **Marie:** She is a volunteer member of the Yvette A. Wright crew who serves largely as the cook. She is a close friend of Carol.
2. **Captain Dodd:** He is the captain of the Yvette A. Wight and an officer in the merchant marine.
3. **The Yvette A. Wight**: This is both the boat and the artificial intelligence that sails it. She is at once a single-masted schooner of some 30 meters, an avatar in virtual space, and a corporate person. She appears on monitors as a young African American woman.
4. **Dr. Carol Delanie**: She is the science officer [SO], an oceanographer who works for Woods Hole.
5. **Tuck**: He is an able-bodied seaman and ex-marine, [ABS1] and an excellent shot.
6. **Carter**: He is a waterman and fisherman, [ABS2] who knows the sea in his gut.

This is the crew of the Yvette A. Wight, a working ship that can broadcast mineral pellets into the ocean (Powell, 2007) and take scientific measurements of the resulting effect. To be effective, the humans in the crew must team with the AI.

Technology

The main cited technologies are: Artificial Intelligence applications and adding minerals to the seas to fertilize plankton growth. For a specific in-depth discussion, see the dedicated Technology Section below which covers such questions as how humans and AIs may function as one crew.

Communicative Approach

There is no doubt that in the communication process the oral language is of primary importance, as well as the starting point for learning a second language.

Stories motivate learning; they are fun and help develop a positive attitude especially towards the second language. The children are involved personally because they often identify with the characters and try to interpret the story and the illustrations thus favoring the development of creativity.

- Listening to stories helps children pay attention to rhythm, intonation, and pronunciation.
- Storytelling facilitates the development of learning strategies such as listening to grasp the general sense, predicting later scenes, guessing meanings, and hypothesizing.
- This story allows the teacher to introduce and revise words and structures, exposing the learners to the language and enriching their linguistic heritage, especially technical keywords.
- In the case of a foreign language, the narration offers opportunities to present cultural information of the country in which the language is spoken. Every word has its own history, its particular meaning, its value that is not always easily translated into another language but that with the right material the teacher can try to "pass" to the students, immersing them in the atmosphere of the second language.

Figure 2. Sunrise viewed across the Chesapeake

Playful Approach

It is therefore important to cut out moments dedicated to activities within the class substantive games and activities, capable of producing true emotional involvement, a fundamental engine of any form of learning.

Transition Into the Story

There are technological solutions to the problems of sea and air. Action is possible. Whatever is done it must be watched very carefully. This will take big data taken from space and air, but it will also take a major hands-on effort.

Sometimes things will go wrong. If a great many people are watching what is being done, when something goes wrong, a great many people will want to argue about what to do next. Is there a story in this effort to get things right? Is there a story in the conflict that occurs when things go wrong?

THE FICTION STORY: "THE CAPTAIN'S TALE" BY TOM RILEY

Premise

Sometimes you are what you do. The sea does not care. The sea is the sea.

At Sea

"Is that damned boat still shadowing us?" asked Captain Dodd while manning the helm in the gathering dawn.

"Yes," replied Yvette, who appeared in her persona as a young African American woman on the large monitor on the aft cabin bulkhead beside him.

"Have you got an ID on it yet?"

"It is the Dawn Light out of Nantucket," replied Yvette, "but it is providing no additional information and apparently does not have an artificial intelligence on board. Also, I have just detected a long-duration drone that has taken position over the harbor entrance ahead."

"Keep an eye on both of them," said the Captain, glancing up at the High-Eye instrument package at the top of Yvette's mast and then behind him at the masts and sails of the distant ship. It was now occasionally visible on the horizon across that gentle sea on this perfect, cool fall morning.

"It is time for the morning crew briefing," said Yvette, "The crew has assembled."

On this particular Iron Seas expedition, the crew consisted of five people and an artificial Intelligence [AI]: Captain Dodd; Dr. Carol Delanie, the science officer [SO, from Woods Hole]; Tuck [an able-bodied seaman and ex-marine, ABS1]; Carter [a waterman and fisherman, ABS2]; and Marie [an eco-tourist temporarily serving as cook]. The Yvette A. Wight was both the boat and the artificial intelligence that sailed it. She was at once a single-masted schooner of some 30 meters, an avatar in virtual space, and a legal person.

"Yvette, take the helm," said the Captain as he moved down into the hatchway where he could see and be seen by the assembled crew. The helm issued a series of bell tones as it shifted to full automatic mode.

"We are a day and a half out of Annapolis, under sail," started Yvette from the large monitor in the main cabin. "The weather looks fine through tomorrow night." A weather map of the mid-Atlantic was displayed behind her.

"If we maintain our planned course and speed, we will make port tomorrow around mid-day. "

"Unfortunately, this is no routine port call," picked up the Captain, "Carol will now provide us with the details of our current mess."

"I am sorry that I did not have time to give you a proper briefing before this," said the Science Officer, Carol. "We are hip deep in a major emergency and headquarters has been on my back, but let me catch you all up now."

"Six weeks ago," she continued, "the Iron Seas bulk ship, the Rachel L. Carson, laid down a major chum grid of 'A' pellets in the Mid-Atlantic off the Chesapeake Bay, some twenty-eight metric tons of mineral pellets went overboard. We were scheduled for a routine science monitoring follow-up when we got word that something had gone terribly wrong. The northern quarter of the field was turning into a red tide. As you are now painfully aware, we have spent the last two weeks sampling that section."

"What a mess," volunteered Tuck, "two weeks of sickly purplish seas that could make you puke mixed with patches of about a billion baby jellies. And dead fish, lots of dead fish."

"Not one hour with the promise of decent fishing in a week," offered Carter.

"Thanks to all your determined efforts," said Carol, "We now have the critical data needed to work out what went wrong. In addition to a large digital data package, a copy of which Yvette has already sent to shore by satellite, we now have actual water and bottom samples in our hold. Solving what went wrong is now my job, but that work can only be done in my big lab on shore. The future of the Iron Seas program now hangs on our data."

"Plenty of people out there want to know what our data mean," picked up the Captain, "and lots of them would be happy to hack the information if it suits their particular politics. It is also now clear that we are being both shadowed from behind and spied on from above. We can certainly expect a media circus when we make port. This dog-and-pony show is all political rubbish, but still our future funding depends on it."

"Marie, as our oldest crew member, could you give us some background on the political angle?" asked Carol.

"To make it short," answered Marie. "When I was young, the eco-politics was between the climate changers and climate deniers. The great mass of people just wanted the problems to go away so they could live their lives in the ways they knew. Then came the big pivot. That one bad summer, back when I was a girl, a lot of people died and the mass of the population did a big turnaround in their attitudes. True or not, they decided that the powers-that-be had been lying to them for a long time. By Christmas, even the conservative cable talking-heads were suddenly born-again tree-huggers. The liberals were at first jubilant but soon started squabbling among themselves over exactly what to do next."

"The official death toll," volunteered Yvette, "was 1.5 million people, 15 thousand of whom were Americans, and twenty-one AIs. That is only counting deaths that were directly attributed to climate change."

"Twenty-one AIs?" someone asked.

"The great fires in the American West took out a major data center," answered Yvette, "much of the data was unrecoverable; also, several AIs that were viable businesses before went bankrupt after."

"I suppose you can die, too; I had not really thought about it," continued Marie. "In any event, politics soon became a contest between the 'Greens' who wanted to take strong action now and the 'Naturals' who wanted to leave nature to heal itself. Just as you would expect, the Iron Seas efforts to sequester the excess atmospheric carbon by seeding the oceans with minerals is loved by the Greens but hated by the Naturals."

"Personally, I am a lifelong Green who still remembers that summer very well," finished Marie.

"There are still a few Deniers around too," added the Captain. "They're a real bunch of nutters."

"One thing is clear," picked up Carol, "If the Iron Seas program is to grow or even continue, then we need to get the data back and find a solution to avoid this type of biological breakdown in the future. Right now, we are being carefully watched by all our opponents, so we must both be open to critical review and protect our work from attack."

"All right then," spoke up the Captain, "The order of the day is making port while looking good for the national media. Tuck and Carter, clean up the last of the mess left by that ugly sea we just took a bath in. Carol, I understand you will be organizing the data. Marie, see if you can work with Yvette and find out which group the Dawn Light belongs to. I will be at the helm most of the day."

Company Coming

Carol came out of the rear hatchway and stood by the captain for a moment. She looked around noting that no one was within earshot for the moment, at least no one except Yvette.

"In a few hours," said Carol, "we will have full bandwidth with the shore. The media will demand that we open an out-only port through Yvette's firewall. Since we get public funding, we cannot really refuse. Every sensor she has will be live on the web and that includes the interior cameras and mics. Our privacy will be gone."

"Yes," answered the Captain, "the fishbowl."

The captain glanced around as the two crewmembers cleaning the aft section worked their way closer. His eyes then fell on the masts of the ship trailing them in the distance.

The moment lost, Carol said quietly, "I will be sure that all my stuff is out of your cabin and all yours is out of mine. I am sure we will have a better chance for a private conversation about where we go from here later."

"Good idea," agreed the Captain, "but it could be much later." His eyes jumped from the trailing ship to Yvette's radar display.

"Captain," called Yvette.

"Yes, what's up?" he asked from the helm.

"Several Items," continued Yvette. "First, there has been a significant attempt to hack my systems. Fortunately, communications are still limited and my firewall held. I think they were after our data."

"Keep me, the science officer, and headquarters immediately informed of all such activities," the Captain responded.

"While we are at sea, our communications are limited and so are their attack opportunities. As we approach shore, that distant protection wains. This attack was relayed through a router on the Dawn Light, but I am sure it did not originate there. Their systems are simply not sophisticated enough."

The captain glanced astern, past the two seamen scrubbing the transom and out to the larger ship now clearly visible. Yvette's great wing mainsail was driving her steadily forward in a freshening breeze but the larger ship could overtake them any time it wished.

"Second, Marie has introduced me to the home AI of the Greens, named Inc. Green America Zero One. It is sure that the Dawn Light is a Natural ship. It says the Greens were organizing a small supporting demonstration to greet us tomorrow but now it thinks the Naturals are planning a much angrier counter rally."

"Then we are in it, a real web storm for sure," agreed the Captain.

"Also, we are getting more than fifty ID inquiries an hour," continued Yvette, "mostly small boats and drones, but our response must include our location for safe navigation. We will certainly have a flock of drones and a flotilla of small boats surrounding us come morning light. The harbor approach is complex enough to make collisions a real possibility."

"Accidentally on purpose," agreed the Captain.

"One other thing, Inc. Green America Zero One is not an independent AI like me. It is designed to be a node in an Earth-wide hive-mind tasked to buttress nature. Think of it as a hill of very smart ants that at the first sign of danger will rush out to protect their home, Mother Earth. They have both smarts and stings. There is clearly a lot more going on than just our small boat entering port. A full-fledged cyber skirmish is a definite possibility."

"Not much we can do about that," said the Captain.

"There is one thing we can do," said Yvette. "GreenAmerica01 has established limited contact with the Nature computer system back at the Natural's headquarters in Boulder. We could try to negotiate a coordinated action tomorrow. Nobody wants to look like the bad guy in front of that much media, so everybody might play nice. I would need your authorization to move forward with this negotiation."

"It might be worth a try," answered the Captain. "Go ahead, but keep headquarters informed."

Once communication was fully established with the shore, the actual negotiation between the three AIs was all electronic and took two minutes and twenty-four seconds.

The Fishbowl

Carol motioned Marie into her very small on-board lab.

"Marie, I know you just heard that the fishbowl will begin soon," said Carol, "Then we will lose all privacy. I need to talk with you before that to help me work out a few of my personal concerns. As you know, I am legally responsible for the chain-of-custody of our data, and once fishbowl begins, that responsibility will be very, very public. I must personally accompany the data all the way to Woods Hole."

Marie entered the lab, which had people space not much larger than a shower, but she did manage to pull the door shut behind her.

"You are the only person I have confided in about Dodd and I being in love," Carol continued, "but what I have not shared with you is that I went off the pill a couple of months ago and could be pregnant."

"Oh, I'm so happy for you; I know your biological clock is ticking!" Marie exclaimed. "I know that this is an important decision that you have considered well."

"He and I, of course, have had serious discussions about what if," said Carol. "We closely agree about most of the issues, especially those concerning the 21st century social mores: the limit on the number of children per couple, when a pregnancy can be aborted and what reasons are acceptable to both of us, and the need to build a virtual village to raise the child. You know them as well as me, if not better. We had expected to have the rest of this voyage as a working honeymoon after which we would go ashore during the boat's scheduled rotation and hopefully confirm that we are expecting and that the fetus is sound. We would then have had enough time to discuss more fully our plans and concerns."

"The algae incident and now the fishbowl means that in a few minutes there can be no private crew conversations either on board or with anybody else anywhere, at least until all the data are safely at Woods Hole. Then I can, hopefully, ditch the press and environmental activists and get to my doctor for the complete tests."

"Still this is wonderful news," replied Marie. "I know how much you want to be a mother and raise a child. The second we go to open data, millions of people on the web will be following us like our lives are some sort of reality show."

"But, you could help me," said Carol. "I want you to help me build a virtual village for the child. In the modern world, it is just too much to expect that a family group will live together for a lifetime. As you know, the best we can do is build a virtual village to help raise the child. Such a group is far more likely to be available electronically during a child's growing period; we can't necessarily expect the parents to be living together after a couple of decades. After all, I will be heavily committed at Woods Hole for at least the next two years over this incident alone. And Dodd, well Dodd is a sea captain and is simply forever looking for a bigger boat."

"I would be honored," answered Marie.

"My hope is that you and Dodd get to know each other better during the rest of this voyage," continued Carol. "Perhaps, then, he will be convinced that you are not here undercover for the Naturals and accept you as a trusted friend of mine and, hopefully, his. You certainly have the experience and maturity to understand many of his concerns and the mood swings that I may be going through as a first-time mother-to-be alone on shore with no family to speak of. After all, my best friends and the baby's father-to-be will be a thousand miles away on this sailboat chumming for CO2 in the ocean!"

"You have all my contact information," said Marie. "I will get busy organizing a proper support group as soon as you know for sure and we have some privacy. We can worry about the formal contract then."

"Thanks so much, but it will be a while," said Carol. "Even after things quieten down it will take a month to get a clean bill of health for the fetus. I don't want to get everybody excited before then."

"It is our secret for now," agreed Marie. "Have you told Dodd?"

"No, the time simply was not right," said Carol. "One other person does know, however, Yvette."

At the sound of her name, Yvette appeared on the small monitor in the lab.

"Marie has agreed to be in our baby's village," said Carol.

Yvette's image started to morph. She changed from an adult to a very young Afro-American schoolgirl dressed in a full skirt. The skirt was pink with a juvenile narwhale embroidered upon it.

"I can provide a special service," said Yvette. "I can guard the child when we are both in virtual space. I am somewhat limited when I am out to sea, but I do have a shore component that can surely do the monitoring even if my avatar is not present."

"Imaginary friends," said Carol, "were once, well, imaginary. Today I guess they are better being virtual!"

"The full monitoring link will be open in two hours," announced Yvette.

"Best back to work then."

Uneasy Lunch

"We now have a coordinated plan for tomorrow," said Yvette from the large monitor over the lunch table.

"This has got to be an AI plan," said Marie, "If it were a people plan, they would still be just talking."

"Most of the plan is simple," continued Yvette. "The Natural boats, including the Dawn Light, will form up in the bay to the north, the Greens to the south. The Coast Guard harbor patrol boat will maintain order. We will approach the harbor along the main channel in a gap in the commercial traffic. We will have a five hundred meter safety keep-out zone around us. All drones must identify themselves and maintain two hundred meters of altitude above all boats."

"Again, at the dock, the Natures have the wharf to the north, Greens to the south. Everybody flies their true banners. Nobody uses dirty tricks, false flags, or crude insults."

"Sounds like an AI plan alright," agreed Tuck. "If there were people involved we would still be arguing about who gets the best camera angles."

"Here's the part you won't like," said the Captain. "As of 14:00 hours today, our full ship's automatic data log will be passed through our firewall and transmitted on an open channel to shore. That applies to all the sensors, including all the on-board cameras and sound. For the next 24 hours, the whole world will be able to see everything that happens on this ship in near real time. "

"Fishbowl time," commented Carter and Tuck simultaneously, marking the phrase with a high-five.

"And the video clips will live on the web forever," added Carol.

"Should I give them my sailor's hornpipe?" asked Carter.

"No," replied the whole crew in unison, including Yvette.

Night at Sea

"Long night," said the Captain to Tuck, who was just completing the last night watch at the helm.

"Feels like a gathering storm," said Tuck, "Only it's made of half-seen people, not bad weather. Still we are holding course and speed. "

"I'll take over now. Take in the foresail," ordered the Captain. "We will make port on the mainsail wing and motor."

A red sun rose at their backs, the sky was clear, and the air had a sharp cut to it. A 25-meter motor launch with the red, white, and blue markings of the Coast Guard now motored 100 meters ahead. A flock of shore birds was winging south across their course in a light breeze.

"The flotillas are assembling," said Yvette. "There has been a steady flow of traffic from all around the bay since last night. Also, a number of news drones is headed our way. Expect a low pass by a big one approaching from the south in about five minutes."

"Prepare to wave and smile," said the Captain.

"Should I launch one of my drones?" asked Yvette.

"No," ordered the Captain, "They are too valuable to your science. The High-Eye is just fine."

A two-meter drone, probably news, approached from the south, altered course, and lined up with the Yvette and the Coast Guard boat. Then it started a slow pass at the minimum sanctioned altitude.

The Coast Guard boat abruptly revved up its two diesels, altered course, and circled around to the north while picking up speed.

"The Coast Guard reports a collision of two small boats and is answering the distress call," said Yvette.

"Aaruuga! Aaruuga! Aaruuga! " Yvette's main alarm sounded.

"A major cyber-attack is in progress," yelled Yvette. "This time they are after control of the boat. I am locking the helm on manual for now. "

"I have the helm. You fight the cyber-attack," barked the Captain. "I can make port manually if necessary."

The alarm then ended abruptly and the helm issued its series of bell tones as it shifted to the locked-in manual mode. Then there was only the sound of the waves lapping the side of the boat for a long five minutes.

"They breached the outer firewall," reported Yvette, "but the Iron Seas information technology people trapped them in a false portico. We fed them a copy of the data set just to keep them busy. Turns out our headquarters IT people have been up all night and were expecting trouble after yesterday's minor event."

"Are you alright?" asked the Captain.

"OK, for now," replied Yvette, "but somebody unknown now knows a lot more about my systems than I am comfortable with. We could put up hard barriers right now but that will block the fishbowl too. It would then look like we are hiding something. Our IT people will meet us at the dock for a rebuild of my whole security system but it will mean an additional day or two of layover in port."

"That doesn't matter," said Carol, who had moved into the hatchway on hearing the alarm. "It will take that extra day or two for my replacement to get here. Headquarters has found a post-doc to complete the cruise but he has to come down from New England."

"Ain't it funny how everything happened all at once?" noted the Captain.

"Funny on purpose more like," said Tuck.

The news drone made another pass and then followed the Coast Guard boat to screen the ramming. The channel markers were now clearly visible in the morning light. The shore was just appearing as a dark green line of trees on the horizon. There were a few patches of gray standing sumps where the salt water had encroached on old forestland. A great bridge with an elegant and distinctive curve loomed in the distance, dressed largely with lines of headlights of evenly spaced robot trucks moving relentlessly at constant speed

The Flotilla

By that morning light, the web wars had begun. Both the Greens and the Naturals offered instant science on the data set. Each side spun its own interpretation to suit its political view. This upset the Woods Hole people who knew they would need a year to properly reduce and then publish the data. For now, the Greens and the Naturals whacked away at each other, leaving the public not knowing what to believe.

Social media dug out the bios of all the crewmembers and yammered about them in excruciating detail. They even went so far as to include the artist after whom Yvette was named. Speculation then ran wild about onboard romances. Fortunately, nothing problematic became public.

For the next two hours, they sailed easily with the captain at the helm. They were accompanied by boats from both political sides, judging by the banners across their sails. The Coast Guard boat resumed the lead position.

Dozens of small drones operated from shore made passes and one or two went into the drink but none came too close. A flock of geese in a graceful V slowly passed them on their way south. The drones only grudgingly granted the bird's air space.

The main marina channel ran at right angles to the wharf. Dodd expertly brought Yvette in, swung her through a right angle, and stopped her dead in the water using only a simple twist of the mainsail wing.

"Make lines fore and aft," ordered the Captain. "Everybody wave to the crowds."

Tuck adroitly stepped onto the dock and made his line fast. Carter did the same near the stern.

The captain spotted an unmarked white van on the dock accompanied by a state highway patrol officer on a big electric motorcycle. The van was parked with it rear doors open, with the license plate blocked from view with a shipping pad. Yvette had skillfully avoided taking a clear picture of the van, plate, or driver.

"There's my ride," noted Carol. "It looks like we'll have a police escort, at least to the state line."

"Carter, Tuck, get the transfer cases out of the hole," barked the Captain. "The sooner we get the boxes out of here, the sooner this media circus packs up and rattles its way out of town."

Following Days

The dawn brought a gray light through a light drizzle.

The clown act had left town though it would follow Carol and the white van for the rest of the day and part of the next. What was left of the crew had spent the night ashore. Yvette waited for better software, hiding behind closed protective walls. The captain waited for supplies and a new science officer.

It was a gray day, perfect for waiting, so the captain spent some time watching the water droplets trail down the cabin windows and slowly drank a cup of coffee. A silenced news window on the cabin monitor showed the white van's progress. They had driven through the night.

A few kilometers to the east, the mighty ocean rolled in a seasonal squall, just as it had for the last four billion years. The small disruptions caused by man will pass as so many other disruptions had rolled past before. What the ocean needed most was time and, perhaps, just perhaps, a little help from its friends.

DIDACTIC ACTIVITIES

The suggested didactic activities focused on several variables as follows:

Key Theme

- *Protection of our ecosystem. Support of artificial intelligence in technical activities at sea.*

Pre-Teaching: It is a pre-reading moment useful for the introduction of ideas and key words present in the story, the verification of the vocabulary already known to the students and useful for understanding the text.

Reading: Descriptions or [if added] images play a key role. At first reading it can focus on the screenplay, postponing it to a second reading the analysis and understanding of the message. For the students, it certainly proves very engaging for them to actively participate in reading by miming some scenes. In this way, the understanding of meaning becomes more immediate and less memorization challenging.

Vocabulary and Several Issues Related to Technology/Science [Quotes as an Example]

- *The crew consisted of five people and an artificial Intelligence [AI]: a Captain, the science officer [SO, from Woods Hole]; an able-bodied seaman and ex-marine, ABS1, a waterman and fisherman, ABS2, the volunteer cook*
- *The Yvette A. Wight was both the boat and the artificial intelligence that sailed it. She was at once a single-masted schooner of some 30 meters, an avatar in virtual space, and a legal person.*
- *sensors, including all the on-board cameras and sound*
- *Foresail*
- *flotillas*
- *cyber-attack*
- *outer firewall*
- *Virtual reality*

Presentation of the Characters: The exercise is impersonating characters [answering questions: What's your name, How old are you ? Where are you from? What are you wearing?], trying to act and interact as the character would have done. The dramatization proves to be a positive and productive activity. Moreover it encourages the student to express himself without inhibitions and allows him to customize a text:

Scenes: The teacher can ask the students to represent the different scenes of the story and write alongside them keywords. The story can be recited and recorded by creating a movie, writing an article for the school newspaper and contributing to the creation of the school website.

Retelling the Story: Students attention is focused entirely on the meaning of story. If understood, they can be encouraged to change some actions or the end of the story.

Skills Involved

- **Comprehension:** Understanding the story; listening to and identifying key words related to interactions between characters.
- **Interaction:** Participate in a recited version of a story through mime and repetition of words.
- **Image Education:** Graphically representing scenes from the story, using different techniques; creation of a comic book graphically depicting a personal reworking of the story.
- **Citizenship Education:** Collaborating within the class group, in a small group and in pairs; consolidating the identity of the class group; reflecting on scientific issues and the technical development of the society.

TECHNICAL DISCUSSION

It Takes a Village to Raise a Child, but What Does This Mean in a Digital Age?

The best available models for world human population for humans (Kurzgesagt, 2016) show a peak of about 10 billion people around 2050. The world population will then settle to some level that the Earth can sustain, but which cannot currently be predicted with any assurance at all. This event is called the Third Demographic Transition (Demographic, 2017)..

The last comparable transition was the industrial revolution about 240 years ago. Just like the last transition, this new transition will have many lasting effects on society. One of which is a reduction in the number of births per woman to below two. This trend is already evident with the developed countries showing the lead. This reduction in the number of births per woman and a general extension of the average person's lifespan will result in many fewer children being around but more people available who are interested in helping to help them as they grow.

In this series of stories, Iron Seas, this value-our-children trend shows up first as the Captain having previously interacted with the new cabin boy by way of an online vetted mentor/prodigy app and as the science officer starting to build a virtual village to support her unborn child throughout its life.

If there are only a few children around then each warrants long-term support from a number of adults, would a virtual village for each child be a good idea? Would having an AI as a member of a child's village be a good idea, particularly to provide protection for the child when in cyberspace?

Figure 3. ICESat-2 image of northern hemisphere ice in winter, NASA (Source: https://www.facebook.com/ICEatNASA/photos/a.111557779038988/9082917393 65584/?type=1&theater)

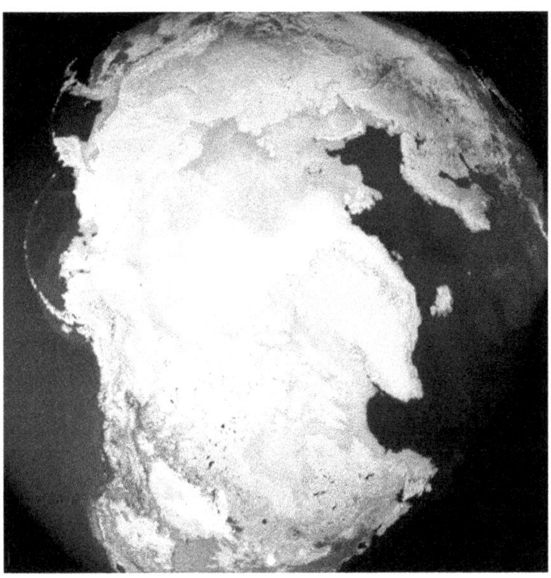

How Is the Land Ice on Greenland Measured?

For many years artic explorers trekked over the Greenland ice on skies and by dog sled but they could not make accurate measurements of the thickness of the ice (Frances, 2018). To determine if the Greenland ice is melting will require many very accurate measurements of the surface elevation year after year.

NASA then built and launched the ICESat [Ice, Cloud, and land Elevation Satellite] mission that provided detailed surface elevation measurements from 2003 to 2009. This mission sent down a powerful laser beam and measured the return time very accurately.

This mission proved so valuable that a continuation mission was immediately planned when the big space laser gave out in 2009. The next generation instrument was designed but building space instruments takes time so an airplane based bridging mission was flown. This mission continued until the new ICESat-2 was launched in the fall of 2018. The Bridge flights will continue until the new instrument is fully operational and thereby provide a transitional link between the two flights.

How important is it to know what the Greenland ice is really doing?

CONCLUSION AND FUTURE RESEARCH DIRECTIONS

The oceans of Earth are vast, even more so than the atmosphere. Society has no silver bullets that would magically restore either to their pristine state. That said, society has knocked both out of long established balances, but people could take actions to restore those balances. The extent of the problem has now been detailed in four major reports (IPBES, 2019; IPCC, 2019; Laybourn-Langton et al., 2019; NCA4, 2018).

Exactly what actions will work best with the least additional disruptions will need to be determined by large scale, but difficult, tests. Sometimes a test will go right and the people closest to the action will learn how to do the task even better. Sometimes the test will go wrong and a critical lesson will be learned at a price.

There are actions that society can take but a much better understanding of the balances is needed to choose the best actions and design big tests. The necessary actions are now limited in scope and disorganized. A clear vision with many organized opportunities for action is possible and clearly necessary.

Transition

The great problems of the 21st century cover both land and sea. There are new management practices for the sea as well as the land. Both can only be worked out by hard work.

The story of the next chapter, "the Matriarch" starts a new series of stories, "The Herd". The biology of both land and sea are filled with symbiosis, where one species lives in mutual corporation with another. Society has broken some of these and must now fix them.

Is there a story to be found in efforts to reestablish broken symbiosis? Is there a story to be found in humans and AIs forming a symbiosis themselves?

REFERENCES

Demographic Transition. (n.d.). In *Wikipedia*. Retrieved February 2017 from https://en.wikipedia.org/wiki/Demographic_transition

Frances, J. A. (2018, April). Meltdown. *Scientific American*, *318*(4), 48–53. doi:10.1038cientificamerican0418-48 PMID:29557956

IPBES. (2019). *Intergovernmental Science-Policy Platform on Biodiversity and Ecosystem Services "Nature's Dangerous Decline 'Unprecedented', Species Extinction Rates 'Accelerating.'"* Retrieved from: https://www.ipbes.net/news/Media-Release-Global-Assessment

IPCC SR1. (2019). *The Intergovernmental Panel on Climate Crisis, Special Report 1.5*. Retrieved from: https://www.ipcc.ch/sr15/

Kurzgesagt – In a Nutshell. (2016, December 16). *Over Population, The Human Explosion Explained* [Video File]. Retrieved from https://www.youtube.com/watch?v=QsBT5EQt348

Laybourn-Langton, L., Rankin, L., & Baxter, D. (2019). *This is a Crisis, Facing up to the age of environmental breakdown*. Retrieved from https://www.ippr.org/files/2019-02/this-is-a-crisis-february2019.pdf

NCA4 Fourth National Climate Assessment. (2018). Retrieved from: https://www.globalchange.gov/nca4

Powell, H. (2007). *Fertilizing the Oceans with Iron, Should we add iron to the sea to help reduce greenhouse gases in the air?* Viewed on April 14, 2018 from: http://www.whoi.edu/oceanus/feature/fertilizing-the-ocean-with-iron

KEY TERMS AND DEFINITIONS

Artificial Intelligence (AI): The intelligence shown by machines or computer software. An AI in some ways can mimic human intelligence but does not have to match it feature for feature. The interface avatars of the AIs are important characters in these stories.

Big Moon Dig (BMD): A fictional grassroots space movement with the purpose of building a settlement on the Moon. It is a MOVE organization. Historically Apollo to the Moon provided an enormous number of people with a vivid, positive vision of the future. Such a vision would be a major asset in addressing the great problems of the 21st century if it can be reestablished. The Big Moon Dig's top-level purpose then is to rebuild that vision.

Iron Seas: A fictional organization that is testing the seeding the oceans with iron to sequester carbon.

Section 3
The Herd Stories

The great problems of the 21st century cover both land and sea. There are new management practices for the land that must be tested on a large scale. Many possible approaches must be worked out by the hard work of doing things by real people.

Science: The key scientific ideas for the Herd series are:

1. **Reestablishing the Symbiosis Between Herd Animals and Grass:** Human actions have let enormous areas of grassland degenerate into deserts. Now large-scale tests have shown that the grasslands can be reestablished by reestablishing the symbiosis of grass and grazing animals. This action will sequester a large amount of carbon in deep rich soils.
2. **Humans and Artificial Intelligence:** The AI in these stories are based on the concept of Human/Machine Symbiosis and not on the more common Master/Slave. This design places a premium on humans and AIs forming teams. These stories show one possible arrangement of the human/AI interaction.

Most ecosystems of the Earth are made up of symbiosis, multiple species supporting each other. For example, most of society's food crops are in symbiosis with the fungus that make up rich soils.

When people damage these symbioses, entire environments can degrade beyond use. This is exactly what has happened in many cases of desertification. The poor use of herd animals has damaged the grasslands where the herds once prospered.

The ruminate animals evolved in symbiosis with grass. The grass evolved its growth buds down in a structure right near the ground. Having their leaves eaten off simply does not hurt the plant all that much. The herd animals evolve big, complex stomachs that could digest cellulose. The hard feet of the animals selectively trampled the other plants, like tree shoots, leaving the grass alive while breaking up hard surfaces. Over time, grasslands, the savannas, became one of the most common ecosystems of Earth. Consider the Serengeti with its great herds, and the central plains of North America with its bison.

Reestablishing the symbiosis of herd animals, in this case cattle, and grasslands has been successfully demonstrated all the way up to some very big ranches. The secret is to restrain the animals to a small patch of grass for a limited amount of time and then move them on. Keeping the herd at just the right density insures that they eat all the available food without wasting very much. Moving them on at just the right speed insures that their urine and waste will water and fertilize the grass. The herds then move in ever-restless great cattle circles.

Once the symbiosis is reestablished, the grass will return even in very dry country. Monitoring this effort on a large scale will not be easy. Big data from satellites will help, but ground truth readings right in the middle of the herd are critical too. This series of stories features an AI controlled all-terrain vehicle, Dusty04, that does most of the monitoring work and a young woman, Zane, whose life is tied up with these cattle circles.

Chapter 7
In Action on Desertification:
A Didactic Introduction

ABSTRACT

Human activities, such as over grazing, have turned grasslands in many places into deserts. New methods in cattle management now allows many of these deserts to be returned to grass. Artificial intelligence is now driving a critical moment in the history of technology. Soon sets of Deep Learning chips will appear both in consumer products and in data centers. This quantum leap in technology will result in AIs that are powerful and cheap, but which use only a little power. This breakthrough will strongly affect the ways humans and machines interact. Our young people need to envision some the ways this transition could happen and to get ready. They need to understand what must be done to keep this expanding technology under human control and how to best use it for the good of society. This chapter's story, "The Matriarch," looks at an interaction between people, animals, and artificial intelligences.

TOPICS FOR DISCUSSION

The following discussion points come from information covered in this chapter:

1. Can the symbiotic relationship between grass and herd animals be used to reverse desertification?
2. Could an autonomous vehicle driven by an AI team with human beings to do work in the field?

DOI: 10.4018/978-1-5225-8401-8.ch007

3. Could an AI that has contact with wild animals aid in improving human communication with intelligent animals like elephants?

INTRODUCTION

Can Artificial Intelligences [AIs] really aid in addressing problems caused by climate change?

According to a World Economic Forum report, "Harnessing Artificial Intelligence for the Earth" (AA.VV, 2018), in the last years, there were more than 800 weather and disaster events, triple the number that occurred in 1980. Twenty percent of species currently face extinction, and that number could rise to 50 percent by 2100. AI refers to computer systems that can sense their environment, think, learn, and act in response to what they sense and their learned objectives.

For example, in India, AI has helped farmers get 30 percent higher groundnut yields per hectare by providing information on preparing the land, applying fertilizer, and choosing sowing dates. In Norway, AI helped create a flexible and autonomous electric grid, integrating more renewable energy.

AI for Earth, a Microsoft programme (Microsoft, 2018) has committed $50 million over five years to create and test new applications for AI. Eventually, it will help scale up and commercialize the most promising projects. Columbia University's Maria Uriarte, a professor of Ecology, Evolution, and Environmental Biology, and Tian Zheng, a statistics professor at the Data Science Institute, received a Microsoft grant to study the effects of Hurricane Maria on the El Yunque National Forest in Puerto Rico. Uriarte and her colleagues want to know how tropical storms, which may worsen with climate change, will affect the distribution of tree species in Puerto Rico.

Another project, named Protection Assistant for Wildlife Security (PAWS) from the University of Southern California (USC, 2018), is using machine learning to predict where poaching may occur in the future. Currently, the algorithm analyzes past ranger patrols and poachers' behavior from crime data; a Microsoft grant will help train it to incorporate real-time data to enable rangers to improve their patrols.

In Washington State, Long Live the Kings (LLTK, 2018) is trying to restore declining steelhead and salmon populations. With a grant from Microsoft, the organization will improve an ecosystem model that gathers data about salmon and steelhead growth, tracks fish and marine mammal movements, and monitors marine conditions. The model will help improve hatchery, harvest, and ecosystem management, and support habitat protection and restoration efforts.

The biologies of both land and sea are filled with symbiosis, where one species lives in mutual corporation with another. When these are broken they can be repaired, but very careful work is needed.

A report presented during the Davos World Economic Forum 2018 (WEF, 2018) highlights more than 80 potential applications of artificial intelligence on an environmental level.

Here are some more possible applications of artificial intelligence that could help save the planet (Vlad et al., 2016; Zahraee et al., 2016).

Autonomous Energy

One of the first uses of AI on an environmental level is the optimization of energy sources, their orchestration based on large-scale consumption forecasts. There are many jobs in this field with systems able to predict, reason, plan, and optimize.

The development of forecasting systems that analyze big data and learn from these predictive models is constantly evolving and has reached a good degree of maturity.

The Department of Energy's SLAC National Accelerator Laboratory (Stanford University, 2017) operated by Stanford University will use machine learning and artificial intelligence to identify vulnerabilities in the grid, strengthen them in advance of failures, and restore power more quickly when failures occur. The system will first study part of the grid in California, analyzing data from renewable power sources, battery storage, and satellite imagery that can show where trees growing over power lines might cause problems in a storm. The goal is to develop a grid that can automatically manage renewable energy without interruption and recover from system failures with little human involvement.

Researchers at the Department of Energy and National Oceanic and Atmospheric Administration [NOAA] (DENO, 2018) are using AIs to better understand atmospheric conditions in order to more accurately project the energy output of wind farms (Cho, 2018).

The estimate is only likely to grow. It means we should finally consider how to make the consumption of energy wiser. It turns out; AI can help with that problem.

Figure 1. Energy consumption from 1990 to 2040 (Source: Kharkovyna, 2018)

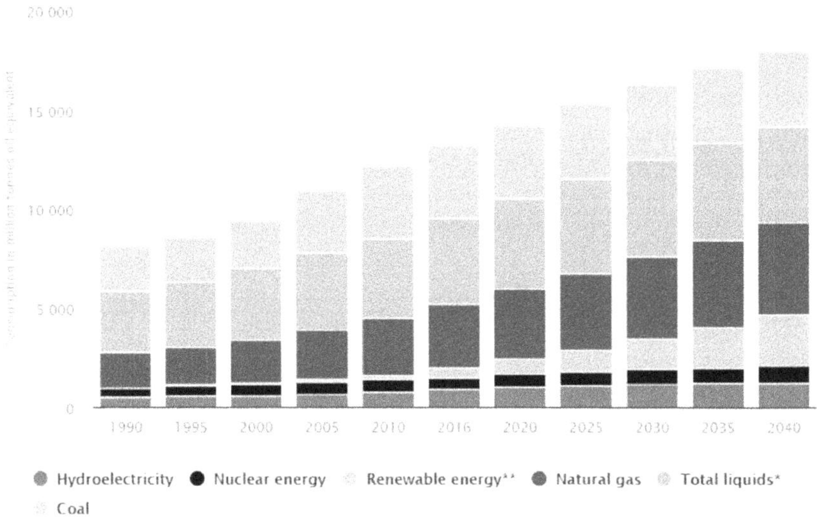

Water Networks

Artificial Intelligence will make a more decentralized water system possible, guided by sensors and new technologies such as *blockchain*, which can allow rapid exchange between activities. Thanks to *blockchain,* it is possible to

Figure 2. Moisture Sensors monitor soil water content for irrigation management (Source: https://commons.wikimedia.org/wiki/File:American_Indian_(38837019171).jpg [09/08/2018])

record the origin of water or energy; the supplier can then stipulate smart contracts and trade between the parts of the decentralized network. The users and producers of the future, whether water or electricity, will be more decentralized and this will improve productivity. Hotter temperatures will have significant impacts on agriculture as well.

Data from sensors in the field that monitor crop moisture, soil composition and temperature help AIs improve production and to know when crops need watering. Incorporating this information with that from drones, which are also used to monitor conditions, can help increasingly automatic AI systems know the best times to plant, spray, and harvest crops, and when to head off diseases and other problems. This will result in increased efficiency, enhanced yields, and lower use of water, fertilizer, and pesticides (Kharkovyna, 2018).

AI Detects Bacteria in the Water

The lack of clean water is one of the few most pressing ecological problems today. Only to test and confirm whether the source of water is clean, one needs s hundred thousand dollars and a lot of expensive equipment.

Figure 3. AI finds patterns of Cholera (Source: https://www.eurekalert.org/pub_releases/2018-05/du-adp050718.php [09/08/2018])

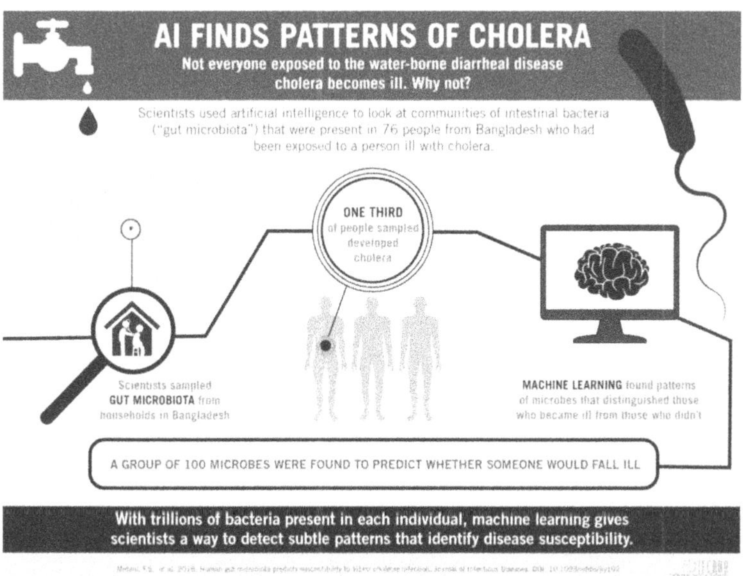

In order to introduce a simpler method of bacteria detection in water, Intel AI academy used artificial intelligence. The team of AI-researchers created a tool for water mapping using machine learning and pattern recognition (Kharkovyna, 2018).

Climate Events Simulations

Climatologists use supercomputers, like that of the Argonne National Laboratory, outside Chicago, to do their calculations. However, there are only a few dozens of real supercomputers in the world, which means that access is limited.

Herweijer (2006) explains that even now, thanks to artificial intelligence, it is possible to make simulations and model the technology on home computers. It is therefore possible to simulate on a small scale some climatic events, such as windstorms. Amazon Third Way is developing a project called Earth Bank of Codes (EBC, 2018), with two main intentions. The first one is open to potential discoveries, the second one foresees to record the origin of the biological resources so that the inhabitants of a location can benefit from the discoveries that derive from it.

Once the artificial intelligence has been integrated into the system, more and more people will be able to carry out simulations and the predictions on the consequences of climate impacts will improve rapidly over the next 10 years. In China, IBM's Green Horizon project (IBM, 2018) is using an AI system that can forecast air pollution, track pollution sources and produce potential strategies to deal with it. It can determine if, for example, it would be more effective to restrict the number of drivers or close certain power plants in order to reduce pollution in a particular area.

Another IBM system in development could help cities plan for future heat waves. AI would simulate the climate at the urban scale and explore different strategies to test how well they ease heat waves. For example, if a city wanted to plant new trees, machine-learning models could determine the best places to plant them to get optimal tree cover and reduce heat from pavement.

AI Underwater

A team of marine robots will revolutionize geophysical underwater explorations. It is a project funded by the European Union, close to commercialization, and aims to revolutionize geophysical and geotechnical underwater explorations.

It is the "Wimust" project (Horizon2020, 2018), which stands for Widely scalable mobile underwater sonar technology, funded by the European Union under the "Horizon 2020" program. The initiative is coordinated by ISME, the inter-university research center on integrated systems for the marine environment (ISNE, 2018) of which the University of Genoa is the registered office. In addition to Italy, researchers from Portugal, the United Kingdom, Germany, France, and the Netherlands participate.

The Ocean Data Alliance is working with machine learning to provide data from satellites and ocean exploration so that decision-makers can monitor shipping, ocean mining, fishing, coral bleaching, or the outbreak of a marine disease. With near real time data, decision-makers and authorities will be able to respond to problems more quickly. Artificial intelligence can also help predict the spread of invasive species, follow marine litter, monitor ocean currents, keep track of dead zones, and measure pollution levels (Grighi, 2018).

AI Monitors Animal Traffic

Species conservation is among the brightest topics on the agenda of environmental societies like the World Wildlife Fund [WWF]. However, society would not need to show so much concern with saving the rare species if society had monitored them more thoroughly (Kharkovyna, 2018).

Sure, it is not an easy job to track every single species in the world; it will require many human resources and a solid funding to back the research up

Figure 4. The H2020 widely scalable mobile underwater sonar technology (Source: Indiveri, 2016)

It was not possible to think of a way to track animals efficiently until now when technology can do many things faster, better, and at a lower cost than with human efforts alone.

How This Chapter Works

The title of this analyzed fiction story is "The Matriarch", the first story of the Herd Series.

This series is about using the symbiosis between herd animals and grass to stop desertification and sequester carbon in the form of organic matter within deep rich soils. There are many grasslands across the Earth that would benefit from this effort and the amount of carbon that could possibly be stored is huge.

Grass evolved with the grazers during the last age of the dinosaurs. Together they form one of the most common ecosystems on Earth. Think of the Serengeti in Africa and the American bison on the Great Plains. The two, grazers and grass, form a symbiosis and neither can prosper alone.

Unfortunately, man interfered with this symbiosis in many places. Deserts spread; carbon in the soil oxidized back into the atmosphere. To reverse these trends will require the application of new management techniques and technologies on a mass scale. Fortunately, a process to do this has been developed and tested largely by Allan Savory (Savory, 2013; Savory & Butterfiel, 2016).

The story is analyzed didactically as fiction storytelling, highlighting how it can be used both to explore the interaction between Nature [Life], Artificial intelligence, and technologies. It addresses ethical issues and provides elements of problem solving management.

Can this story tell of how this approach might turn into a mass movement and tell it well enough to help make that mass movement happen? Is there a story to be found in efforts to reestablish broken symbiosis? Is there a story to be found in humans and AIs forming a symbiosis together to take on great problems?

DIDACTIC ANALYSIS OF "THE MATRIARCH"

The story is analyzed as science fiction storytelling, highlighting how it can be used both to explore crucial science issues on addressing desertification and improve computational thinking and creative problem-solving.

Environments

The scene takes place in a sunny countryside on the African Sahel at a small power station.

Events and Actions

The story describes both natural events and resource management by people and AI machines. The primary event is the encounter between an AI and an elephant.

Main Characters

The main characters in this story are:

1. **Dusty04**: This is a self-driving off road vehicle and corporate person with an AI. It has responsibilities for managing a herd of cattle.
2. **Abdalla:** He is an African cattleman and herd supervisor. He works with Dusty04.
3. **The Matriarch:** She is a large female elephant who is the leader of her family group.
4. **JanetA:** She is a major AI and one of the team of Janet sisters who support all the non-governmental (NGO) projects that address problems of global warming. Her specific responsibilities are to supervise all the Savory cattle circles in Africa. The Janets do not have physical robotic presences but can appear in any number of electronic monitors. JanetA appears as an African woman.
5. **Zane:** She is a young African girl. Scenes from her life story are woven into the many of stories in the Herd series. Her life is strongly effected by her interactions with the Janets.
6. **Imani:** She is Zane's mother.

These are people whose lives are built around the Savory effort to use cattle to stop the advance of the deserts. Their interactions with the AI on this project greatly effect their lives.

Technology

The main cited technologies are: Artificial Intelligence machines, a water control system, a watering hole, a pump. For a specific in-depth discussion, see the dedicated Technology Section below which includes supporting topics.

Communicative Approach

Stories motivate learning. Students are involved personally because they often identify with the characters [or animals] and try to interpret the story and the illustrations thus favoring the development of creativity.

- Listening to stories helps Students pay attention to rhythm, intonation, and pronunciation.
- Storytelling facilitates the development of learning strategies such as listening to grasp the general sense, predicting later scenes, guessing meanings, and hypothesizing.
- This story allows the teacher to introduce and revise words and structures, exposing the learners to the language and enriching their linguistic heritage, especially technical keywords.
- In the specific case of a foreign language, the narration offers opportunities to present cultural information of the country in which the language is spoken. Every word has its own history, its particular meaning, its value that is not always easily translated into another language but that with the right material we can try to "pass" to the students, immersing them in the atmosphere of the second language.

Playful Approach

It is therefore important to cut out moments dedicated to activities within the class for substantial games and activities, capable of producing true emotional involvement, a fundamental engine of any form of learning.

Transition Into the Story

The great problems of the 21st century cover both land and sea. There are new management practices for the land as well as the sea that must be tested on a large scale. Both must be worked out by the hard work of doing.

Figure 5. The Matriarch

The biologies of both land and sea are filled with symbiosis, where one species lives in mutual corporation with another. When these are broken they can be repaired, but very careful work is needed.

Is there a story to be found in efforts to reestablish broken symbiosis? Is there a story to be found in humans and AIs forming a symbiosis together to take on great problems?

THE FICTION STORY: "THE MATRIARCH"

Premise

If you do not change your ways today,
Then with us you will ride,
Trying to catch that devil's herd across these endless skies.
-- Ghost Riders in the Sky

At the Pump

"When will the herd reach this watering place?" asked Dusty04. The AI's built-in cell phone read only one bar, but the signal was steady.

"Tomorrow about sunset," replied Abdalla. "They're a thirsty lot already, but they can't smell the water just yet." The sun was hot in a nearly cloudless sky, and both knew that sun was what now drove the cattle.

"This watering hole is in a good state," continued Dusty04, "the batteries are charged, wind and solar are both producing power, the farmer's fields are being opened up now, and the water level in the well is above the minimum required."

"I will keep the herd rolling," said Abdalla.

"The local villagers were warned last market day, but they are still out moving their electric fences to keep the cattle out of the village," continued Dusty04. "I will start the pump for all the well can take just before dark."

"We saw some lion sign yesterday," said Abdalla. "That should keep the herd bunched and moving."

"Wait a minute," injected Dusty04, "we have company."

A large female elephant moved out of the bush 50 meters to the north, pausing to survey the small power station and pump head. She cast a weary eye on Dusty04, who was parked at the charging station by the solar panels. She extended her trunk toward the long, low trough that led from the pump head out to the now nearly dry stock pond. She could smell the remnants of water left from the villagers drawing their day's needs but quickly understood that there was too little for her family to drink, and certainly not enough for a proper mud bath.

The stock pond was where the herd must drink this time tomorrow. Dusty04 planned to run the pump all night and the cattle would need water desperately come dawn.

"Take care," warned Abdalla, "We cannot stand any elephant damage to the pump site with the herd this close."

Dusty04, a self-driving off-road vehicle with only a modest AI, made no sound and did not move. It dropped Abdalla's cell line and opened one up to the much heavier AI back at headquarters, JanetA, its boss; it called in the heavy guns.

The elephant made a low rumbling sound clearly intended for her family still hidden in the dry bush. The message's frequency band was far below anything a human could hear but Dusty04 was built to handle it. He transmitted the sounds to JanetA who put her heavy speech recognition routines and some deep learning circuits on it.

"I have the words 'family', 'water', and 'people'," said JanetA in the buzzy language that AIs use to talk to each other much more quickly than they can to humans.

"All she wants is a drink of water," said Dusty04, "I will see if I can give her a drink and then ask her to move on."

“We have only twenty-three reliable phrases with gestures for elephant on record,” said JanetA. “That is a good start, but that might not be enough. I don’t see a better plan, so this looks like a good time to learn a few more words of elephant.”

Any great herd circuit, and now there are quite a few up and running, requires a watering hole at no less than every three days’ walk by the cattle. Most of these are natural watercourses, but three wells had to be drilled to allow this particular circuit to function. This was typical for many others spread across the African Sahel. Drilling deep wells in remote locations is very expensive, but they could now be financed with international carbon credits and crowd funding.

These wells produce only limited water in this long-dry country, too little for commercial irrigation, but there was enough for the daily needs of a village, their kitchen gardens, and some small livestock, mostly chickens. Much of the well’s aquifer had been deposited ten thousand years ago, the last time this land was green and lush. A few times a year, depending on the recovery rate of the local grass, these wells were stripped to satisfy a passing herd. Of course, the smell of water attracted wild life, but having a wellhead trashed just now would be a major disaster.

JanetA sent out an Action Alert to all the people in her amateur science group for Africa. It would take several minutes for even the fastest of these people to login and form a support group, by then the critical action would probably be over.

Dusty04 sent several photos of the elephant to JanetA. The elephant was not wearing a radio collar, but the science support team might be able to identify her from her ear contours.

Dusty04 then downloaded all the words known to be used by elephants. The sound files were stored in a form that could be played back at ten times speed so humans could hear them; JanetA rebuilt the audio files to the original super-low tones on the fly.

“Great Matriarch!” called Dusty04 as loudly as he could in elephant. This stressed its sub-woofer past its limit and it started to overheat but it held. Dusty04 rolled forward a meter, flashed its headlights, and did a bounce with his front suspension, just to be sure the elephant could see who was speaking.

The elephant froze, stared at Dusty04 for half a second, then charged forward ten meters, stopped suddenly, bellowed loudly, and shook her great head.

“You *not* my family!” she commanded as a threat.

The elephant was large for a female and of a light dusty-gray color, but had only small tusks, and the tip of one was broken. Her poor ivory had probably saved her life more than once. Was the scar on her left flank an old bullet wound? The rest of her family were now moving up behind her but they remained surprisingly well hidden in the rough bush.

"I got 'not' and 'family' again," transmitted JanetA, "The tone is definitely a challenge."

The Matriarch's challenge to Dusty04 was indicated by a tonal slide in the low registers. No human could make it out, but JanetA could, just barely.

"True," returned Dusty04. "I guard the people-buffalos."

"People-buffalos" was a known word for cattle.

"You guard water, people-rhinoceros!" demanded the Matriarch.

"'People-rhinoceros' clearly means you," transmitted JanetA.

The term "people-rhinoceros" was known to be the elephant word for the Land Rovers that carry tourists and filmmakers. Dusty04 found this description of itself acceptable.

Dusty04 had the body of a mini all-electric flatbed truck, but most of the cab was replaced with a 3D-printed shell full of electronics. It did have arms making it a true robot. They were long and spindly ones but strong; right now, both were stowed along its side, tucked under its truck bed. Its truck bed was just big enough to carry one of the four-wheel all-terrain vehicles use by the wranglers, or occasionally, one dead cow. Dusty04 could do almost every task needed on a Holistic Herd Circuit, and the few it could not do, like castrating the calves to make them into steers, it could teach the locals how to do. In the field, Dusty04 was in charge and fortunately, it did not cost quite as much as a well.

"People-buffaloes" was clearly the Matriarch's name for cattle.

"You guard water!" insisted the Matriarch again. Her ears were now out and her head held high. The intense gaze of her eye alone would stop you cold in your tracks.

"I wait to give water to the people-buffalos," answered Dusty04 as best he could.

"Give family water!" demanded the Matriarch. She snorted and stomped the dry ground with her front feet.

This put Dusty04 in a bind. Watering this single family of elephants one time would do no harm. In fact, elephants were needed to keep the restored grasslands from turning into forests. Nevertheless, another day the elephants might return, not find him, and trash the pumping station looking for water again. They were certainly big and strong enough to do real damage.

"Show you dig water here," said Dusty04. "I call people-matriarch."

It seems that elephants have many phrases about water and even more about their families. Dusty04 phoned Imani, the head of the women's club for the local village, and spoke for only a few moments. It had to put JanetA on hold momentarily to do so.

The Matriarch called out a name in a tone you felt more than heard, and a young female elephant moved forward and stood beside her great aunt. The Matriarch then waited uneasily, sniffing the air and exchanging low tones with her family. The smell of people was too strong here for any elephant to stand easy.

A brown-and-tan bird with a bright orange bill landed on the Matriarch's back. It jumped around, turning its head this way and that, looking for ticks. Several species of birds were critical to the herd cycle. The egrets and larger birds worked among the cattle eating any insects the cows startled. They were critical to controlling locusts and other crop pests. The oxpeckers likewise controlled ticks right off the cows' backs. Even if they ate more cow than insect, they kept down disease. The birds also spooked easily, giving the herd warning of any danger.

A few moments later, Imani came walking up the path from the village holding her cell phone and with her daughter Zane following closely but hiding behind her skirt. Then Imani saw the elephant, stopped with a shocked face, and raised her cell phone.

"What are those elephants doing here!" Imani demanded.

"They only want a drink," said Dusty04. "I'm sure they'll leave after they've had some water."

Dusty04 also knew that the villagers' electric fences would not stop elephants. Electric fences just make elephants mad. They rip them out. It was dangerous to the village even to have elephants around.

"But what if they make their home nearby and start to eat my village's crops?" Imani insisted.

Imani took ten steps forward and set her feet directly facing the Matriarch. Dusty04 translated the message but with much less of a challenge. The Matriarch once more bellowed, stomped her front feet, and stared Imani down.

"But we need the elephants for the herd cycle," said Dusty04, "They keep the grasslands from turning into a forest." He was now speaking in as conciliatory a tone as he could manage while speaking in two such dissimilar languages.

The Matriarch shifted her gaze to Dusty when he spoke.

"Well, just be sure they know whose home this is!" Imani said, also turning toward Dusty and thus breaking the standoff.

"Please tell Zane to push the Big Red Button to demonstrate for the elephants how the water system works," picked up Dusty04.

Imani motioned with her hand while turning her eyes back to the Matriarch. The small girl scampered boldly up to the metal box beside the wellhead, just as she did every day to draw her family's water. She reached through an opening in the protective fence and pressed the Big Red Button.

The Matriarch was startled by the low growling sounds now coming from deep below the ground. Water gushed forth, first into the catch basin designed for filling water jugs, and then overflowed into the long animal trough. The smell of water permeated the dry air. The dry mud at the bottom of the trough darkened in color. Nevertheless, after a couple minutes, before the water running down the trough actually reached the elephants, the flow stopped abruptly.

"Press the Big Red Button again," instructed Dusty04.

Imani waved again. The girl complied, the water flowed again, and this time, most of it went down the animal trough.

"Thank you, it is safe to go now," said Dusty04. The girl ran quickly back down the path in the direction of the village followed by her mother, who was now holding her cell phone tightly against her chest. Dusty04 was sure it heard the word "elephants" and the phrase "talking truck" several times in a very despairing tone through the too fast beat of her heart.

The Matriarch spoke again but Dusty04 could not make out very much of it. JanetA's language-learning routines were running at maximum, but there was only a small established foundation of elephant phrases to work from.

The young elephant slowly approached the wellhead control box, extended her trunk, and carefully smelled everything. Only then did she push the Big Red Button with her trunk finger.

The water flowed again. The Matriarch moved forward to smell and taste it. She then backed away and called something out, again in low tones. Her family moved forward and drank. Soon the young female was gaily playing with the Big Red Button, prancing around, sloppily drinking from the basin, and flapping her ears.

When all had drunk, the Matriarch turned to Dusty04. She then bellowed her full name and her title as matriarch. The title was well understood, but her name had no reasonable translation in any human tongue.

"Who you?" demanded the Matriarch.

The Artificial Intelligence's name, "Inc. Dusty, Dusty, Dust-Face, Zero Four," did not translate easily into elephant.

"Dust-Face," answered Dusty04. It was the best anyone could have done.

“No,” said the Matriarch. “You are People-Rhinoceros-Find-Water.”

Taking no further notice of Dusty04, the Matriarch turned and led her family away, and they were soon lost in the bush. Lions would smell the dampness left in the trough come night; her family must be far away by then. Besides, the people smell was much too strong here.

Reporting In

Dusty04 wondered if every elephant in the district would know its new name, “People-Rhinoceros-Find-Water,” by the time the rains came, or if the Matriarch would keep it as a family secret.

Would the Big Red Button be discussed in low tones that carried across the savanna? Would the elephant families speak of water at the people place? Would other elephant families come? Time would tell.

Dusty04 wondered if the elephants understood that the great circuit of the herd had brought back the grass that now sustained her and her family, and that it had stopped the advance of the great desert that had been pushing south for generations of both people and elephants. Perhaps they never would know; perhaps they already knew.

Dusty04 filed a report of the sound recordings with the science team. It was sure that there were new elephant words in these files, but it could, of course, be wrong and be reading too much into the whole incident.

The wellhead controller had also logged everything that had happened near it, including images of anything that moved, but it did not have infrasound capabilities or an AI, and so it could only log and then transmit the raw data.

Dusty04 knew that as dawn rolled around the Earth, waking up the big cities, thousands of citizen scientists would download the new elephant language files in a great wave. Tens of thousands of people supported the Holistic Cycles by putting eyeballs on its data. Their biggest contribution to date was identifying crocked politicians, rustlers, and other wastrels, but what they loved most was new science. Dusty04 was very sure that they would eat the new files alive, and before the files were properly digested, the nature documentary people would come running out into the field to interview Dusty04 itself. Nevertheless, that probably would not happen until after the rains had come and gone.

Despite what the wranglers or the villagers might think, Dusty04 was not the smarts of this outfit; the big AI far away was; she was called JanetA. Every part of the herd circuit’s ecosystem had to be managed and managed well for the Holistic Circuit to work at all. The big computer did all the forecasts that made this level of management even possible. If satellite measurements were necessary, it could get them. If international networking was needed, it could do just that. Dusty04 was its eyes and ears in the field; Dusty04 provided ground-truth. Dusty04 could show just how much the cows had grazed off the grass, just how much the cows' hooves had broken up the hard ground surface, just how evenly the dung and urine were distributed, and just what plants and animals were thriving and which were disappearing. Some days Dusty04 actually was able to do some science.

"Abdalla, the elephants are gone," said Dusty04, ringing back. "And the water is now flowing, and I have turned over what I learned of elephant to the science team."

"Good," replied Abdalla, who had been following the action intermittently on his phone and by viewing the pump head images. "I'd like to be there when the nature camera people come out and try to interview the Matriarch, what a hoot!"

Abdalla thought about it for a minute and then guessed he would be nowhere near when that happened. He would be on his four-wheeler moving the herd. He would be riding screen between the herd and the lions. He would be riding drag for days, bringing up the rear, pushing stragglers, and eating dust. He was a wrangler; it was a good job in this country even if you worked for a talking truck.

Every now and then, Abdalla even got into some town they passed for a beer and some music. On a good day, the African rhythms spoke to his soul. On a bad day, he was just left with a hangover. A couple times a year he would go back to his home village and stay for a while when a stationary crew were processing the herd in a larger railhead town or when the seasonal rains made movement impossible.

The Matriarch now knew that the people-buffalo herd was coming. She knew just how big it had become and just how quickly it had grown since she first saw it as a calf. She knew that the herd would eat the best plants in the open places, tear up the hard surface of any exposed soil, and leave behind stubble fouled with their urine and dung. She also knew that the grass would then grow back lush and full after the rains. She knew that the lions dare not take on the herd directly, but they would be on the hunt for stragglers. The lions would keep the herd bunched and in the open, out of the bush. It was in that bush she could keep her family safely hidden and fed until the herd passed. She was not afraid of the lions; she had fought off lions before. Besides, she now knew how to dig water in this people-buffalo watering place. That she would never forget; that is what it is to be a matriarch.

Dusty04 finished its charge and checked the pumping rate for the cattle. If the wind held its present direction, they would smell the water by dawn. It would take everything this well had to satisfy the herd for the next three days, then they would all move on; the herd always moves on.

Dusty04 went to town itself twice a year or so for maintenance and upgrades; it was just beginning to appreciate the music there. Beyond that, the herd was its life, and the people and machines that serviced the herd were its village.

The sun was now low and the sky still spoke of dust much more than of the long-promised rains. Now fully charged, Dusty04 rolled off toward the village, there to facilitate an evening village council meeting and then to teach a class.

DIDACTIC ACTIVITIES

The suggested didactic activities focused on several variables as follows.

Key Themes

- The biologies of both land and sea are filled with symbiosis, where one species lives in mutual corporation with another.
- When these are broken they can be repaired, using high technology

Pre-Teaching: It is a pre-reading moment useful for the introduction of ideas and words key present in the story, the verification of the vocabulary already known to the students and useful for understanding the text.

Reading: Descriptions or (if added) images play a key role. At first reading it can focus on the screenplay, postponing it to a second reading the analysis and understanding of the message. For the students, it certainly proves very engaging for them actively participate in reading by miming some scenes. In this way, the understanding of meaning becomes more immediate and less memorization challenging.

Vocabulary and Several Issues Related to Environments and Nature [Quotes as an Example]

- *"The sun was hot in a nearly cloudless sky, and both knew that sun was what now drove the cattle."*
- *"A large female elephant [...] could smell the remnants of water left from the villagers drawing their day's needs but quickly understood that there was too little for her family to drink and certainly not enough for a proper mud bath".*
- *"Much of the well's aquifer had been deposited ten thousand years ago, the last time this land was green and lush."*
- *"The elephant [...] bellowed loudly, and shook her great head"*
- *"The elephant was large for a female and of a light dusty-gray color, but had only small tusks, and the tip of one was broken."*
- *"A brown-and-tan bird with a bright orange bill"*
- *"It jumped around, turning its head this way and that, looking for ticks."*
- *"Several species of birds were critical to the herd cycle."*
- *"The egrets and larger birds worked among the cattle eating any insects the cows startled."*
- *"They were critical to controlling locusts and other crop pests."*
- *"The oxpeckers likewise controlled ticks right off the cows' backs."*
- *"Even if they ate more cow than insect, they kept down disease."*
- *"The birds also spooked easily, giving the herd warning of any danger".*
- *"we need the elephants for the herd cycle, [...] They keep the grasslands from turning into a forest."*
- *"Water gushed forth, first into the catch basin designed for filling water jugs, and then overflowed into the long animal trough".*
- *"The smell of water permeated the dry air".*
- *"The dry mud at the bottom of the trough darkened in color".*
- *"Lions would smell the dampness left in the trough come night"*
- *"The sun was now low and the sky still spoke of dust much more than of the long-promised rains".*

Vocabulary and Several Issues Related to Technology/Science [Quotes as an Example]

- *Artificial Intelligence*
- *electric fences*

- *solar panels*
- *a self-driving off-road vehicle with only a modest AI*
- *"The message's frequency band was far below anything a human could hear"*
- *"heavy speech recognition routines and some deep learning circuits on it"*
- *"[AI machines, in the story,] have only twenty-three reliable phrases with gestures for elephant on record"*
- *"[there are] natural water courses, but three wells had to be drilled to allow this particular circuit to function"*
- *"Drilling deep wells in remote locations is very expensive, but they could now be financed with international carbon credits and crowd-funding"*
- *"These wells produce only limited water in this long-dry country, too little for commercial irrigation, but there was enough for the daily needs of a village, their kitchen gardens, and some small livestock, mostly chickens".*
- *sub-woofer*
- *mini all-electric flatbed truck*
- *3D-printing*
- *"[AI machine's] truck bed was just big enough to carry one of the four-wheel all-terrain vehicles use by the wranglers, or occasionally, one dead cow"*
- *water system*
- *wellhead controller*
- *infrasound*
- *"The wellhead controller had also logged everything that had happened near it, including images of anything that moved, but it did not have infrasound capabilities or an AI, so it could only log and then transmit the raw data."*
- *"Every part of the herd circuit's ecosystem had to be managed and managed well for the Holistic Circuit to work at all"*

Social Issues: *The Human/Machine Symbiosis,* one of the key themes.
Quotes:

- *"Every now and then, [Abdalla] even got into some town they passed for a beer and some music".*
- *"On a good day, the African rhythms spoke to his soul".*
- *"a stationary crew were processing the herd in a larger railhead town or when the seasonal rains made movement impossible".*

Presentation of the Characters: The exercise is impersonating characters [answering questions: What's your name, How old are you ? Where are you from? What are you wearing?], trying to act and interact as the character would have done. The dramatization proves to be a positive and productive activity. Moreover it encourages the student to express himself without inhibitions and allows him to customize a text:

Scenes: The teacher can ask the students to represent the different scenes of the story and write alongside them keyword. The story can be recited and recorded by creating a movie, writing an article for the school newspaper and contributing to the creation of the school website.

Retelling the Story: Students attention is focused entirely on the meaning of story. If understood, they can be encouraged to change some actions or the same final of the story.

Skills Involved

- **Comprehension:** Understanding the story; listening to and identifying key words related to interactions between characters.
- **Interaction:** Participate in a recited version of a story through mime and repetition of words.
- **Image Education:** Graphically representing scenes from the story, using different techniques; creation of a comic book of personal reworking of the story.
- **Citizenship Education:** Collaborating within the class group, in a small group and in pairs; consolidating the identity of the class group; reflecting on scientific issues and the technical development of the society.

TECHNICAL DISCUSSION

If We Built Very Big AIs to Take on Very Big Problems, What Would They Be Like?

To drive a car, an AI must predict the future every few milliseconds. It must answer the question "what would a good driver do next?" With the coming introduction of specialty chips for deep learning, it may be possible to build a big AI to, every few minutes, answer the question, "what should my team do next? Such an AI would also have access to an enormous amount of big data. In these stories, these big AIs are:

Janets: In these stories, the Janets are a series of fictional high-end AIs that are specially designed to manage major technical projects intended to address the great problems of the 21st century [global warming, sea level rise, etc.]. All the units in this series have female interface avatars so they are addressed as female.

In these stories, the major series of AI are given interface avatars taken from either fiction or history. This reduces the task of the designers who would otherwise have to invent something completely new and makes the personalities easier for other people to interface. The Janets are based a character from Australian television, Janet King, who was played brilliantly by Marta Dusseldorp. This series has appeared on Public Television in the United States.

The foundation of their design is the Human/Machine Symbiosis and definitely **not** Master/Slave.

What Is a Symbiosis?

Very commonly in biology, two quite different species evolve to support each other (Symbiosis, 2018). In extreme cases, neither can reproduce without the help of the other; in less extreme cases, the two can survive independently but only thrive with together. This relationship is called symbiosis.

Figure 6. Field in Mexico before planned grazing (Source: Savory, 2013).

Figure 7. Field in Mexico with planned grazing (Source: Savory, 2013).

In these stories, symbioses is critical both to the relationship between humans and their machines and to the human's efforts to address great ecological problems. Cattle and grass evolved together in symbiosis but are now in trouble and require careful management by humans to regain their critical place in nature. The oceans microorganisms can maintain the level of carbon dioxide in the atmosphere optimal for current life but now require careful human management to address problems caused by humans.

How critical is understanding symbiosis to addressing the great problems of the 21st century?

What Will Happen When an AI Meets Wild Life?

In this story, "The Matriarch", a moderately smart instrument carrier, Dusty04, is placed in the field. Its key purpose is to very carefully measure the effects the herd of cattle is having on the local environment. It works closely with a larger, fixed Artificial Intelligence, JanetA, that is trained specifically to manage this type of operation over a large area.

To do its job, Dusty04, has a number of instruments to carefully measure the entire environment including the wild life it encounters. It also can make all the sounds used by the humans in its area in the full range from whistles, through talk, to drum beats.

JanetA has a significant language capability as she must deal with many human beings from many regions. She also has substantial deep learning capabilities as well as access to the latest scientific research.

The story line here simply asks what would happen if they encounter a non-human yet intelligent being, in this case an elephant. This fictional encounter makes a good story, but, what would really happen?

CONCLUSION AND FUTURE RESEARCH DIRECTIONS

These stories are based on the idea that humans and their machines will exist in a Human/Machine Symbiosis. This arrangement is the concept of symbiosis so common in nature, but has not yet been strongly implemented in computer code.

The currently favored foundation is Master/Slave. The major advantages of Master/Slave are that it is easy to implement and is attractive to the master. Therefore, it is currently the most common arrangement by a wide margin. The disabling problem with Master/Slave is that it is practically impossible to write code where the master and slave cannot switch. Being the master may be acceptable at first, but one day suddenly finding oneself the slave is not acceptable at all.

In these stories, the major Artificial Intelligences are independently incorporated. In the United States, this simple legal process is cheap and currently widely available. This status then gives them some rights as a person (Corporate Person) ranging from the right to have a bank account to the right to hold religious beliefs. This proposal is often described as a cheap trick. It certainly needs to be tested and evaluated more fully. In the end, its wide use might even require a court case.

Incorporation also makes the AI an economic entity that has limiting monetary liabilities that could result in the AI going bankrupt. All the AIs in these stories are therefore in their own way mortal. The effect of mortality on an AI warrants much farther discussion.

These stories call for a Committee for the Protection of the Human/Machine Symbiosis to manage the overall human/machine interface. Society clearly needs to set some rules before for a powerful AI goes too far. Could such a committee handle this function? To whom would the committee be responsible?

Transition

If an AI can be a person, how about an elephant, or a whale, or a bonobo? Who sets this rule?

If AIs and people are working together in symbiosis, who is in charge? Do they take turns? Should there be a ruling council of some sort?

The story in the next chapter, "The Triceratops", considers a situation where AIs are both training young people and making decisions that will strongly effect the course of their lives. If the AI see a need for trained people in the future, how much power should it have over the training of people? Would you accept career advice from an AI? Is there a story in who is a person? Is there a story in AIs training people?

REFERENCES

AA.VV. (2018). *Harnessing Artificial Intelligence for the Earth*. Retrieved from: http://www3.weforum.org/docs/Harnessing_Artificial_Intelligence_for_the_Earth_report_2018.pdf [09/08/2018]

Cho, R. (2018). *Artificial intelligence—a game changer for climate change and the environment*. State of the Planet. Retrieved from: https://phys.org/news/2018-06-artificial-intelligencea-game-changer-climate.html

DENO. (2018). *Earth System Research Laboratory*. Department of Energy and National Oceanic and Atmospheric Administration (NOAA). Retrieved from: https://www.esrl.noaa.gov/about/ [09/08/2018]

EBC Earth Bank of Codes. (2018). *Making nature's biological assets visible and valuable*. Retrieved from: https://www.earthbankofcodes.org/

Grighi, L. (2018), *I 5 modi in cui l'intelligenza artificiale può aiutare a salvare il pianeta*. Retrieved from: http://startupitalia.eu/86405-20180228-5-modi-cui-lintelligenza-artificiale-puo-aiutare-salvare-pianeta

Herweijer, C., Seager, R., & Cook, E. R. (2006). North American droughts of the mid to late nineteenth century: A history, simulation, and implication for mediaeval drought. *The Holocene*, *16*(2), 159–171. doi:10.1191/0959683606hl917rp

Horizon2020. (2018). *WiMUST (Widely scalable Mobile Underwater Sonar Technology)*. Retrieved from: http://www.wimust.eu/

IBM. (2018). *IBM's Green Horizon project*. Retrieved from: https://www.research.ibm.com/green-horizons/interactive/

Indiveri, G. (2016). *The H2020 widely scalable mobile underwater sonar technology (WIMUST) project: An overview*. Retrieved from: http://sidra2016.dia.uniroma3.it/h2020-widely-scalable-mobile-underwater-sonar-technology-wimust-project-overview/

ISNE. (2018). Retrieved from: http://www.isme.unige.it/

Kharkovyna, O. (2018). *How Is AI Saving The Planet*. Retrieved from: https://becominghuman.ai/how-is-ai-saving-the-planet-92473b41cfa0

LLTK. (2018). *Long Live the Kings*. Retrieved from: https://lltk.org/

Microsoft. (2018). *AI for Earth*. Retrieved from: https://www.microsoft.com/en-us/aiforearth

Savory, A. (2013). How to fight desertification and reverse climate change. *TED*. Retrieved from: https://www.ted.com/talks/allan_savory_how_to_green_the_world_s_deserts_and_reverse_climate_change?language=en

Savory, A., & Butterfield, J. (2016). *Holistic Management, A Common Sense Revolution to Restore our Environment*. Island Press.

Stanford University. (2017). *SLAC-Led Project Will Use Artificial Intelligence to Prevent or Minimize Electric Grid Failures 2017*. Retrieved from: https://www6.slac.stanford.edu/news/2017-09-14-slac-led-project-will-use-artificial-intelligence-prevent-or-minimize-electric-grid

USC. (2018). *Protection Assistant for Wildlife Security (PAWS)*. University of Southern California. Retrieved from: https://www.cais.usc.edu/projects/wildlife-security/

Vlad, C., Barbu, M., & Vilanova, R. (2016). Intelligent Control of a Distributed Energy Generation System Based on Renewable Sources. Sustainability, 8, 748 doi:10.3390u8080748

WEF. (2018). *The Davos World Economic Forum 2018*. Retrieved from: https://www.weforum.org/events/world-economic-forum-annual-meeting-2018

Zahraee, S. M., Khalaji Assadi, M., & Saidur, R. (2016). Application of Artificial Intelligence Methods for Hybrid Energy System Optimization. *Renewable & Sustainable Energy Reviews*, *66*, 617–630. doi:10.1016/j.rser.2016.08.028

KEY TERMS AND DEFINITIONS

Artificial Intelligence (AI): The intelligence shown by machines or computer software. An AI in some ways can mimics human intelligence but does not have to match it feature for feature. The interface avatars of these AI are important characters in these stories.

Corporate Person: A series of powers and responsibilities that are given to corporations that allow them many of the rights of a real person.

Ground Truth: Measurements made on the ground by instruments that can be regularly calculated. Such measurements provide a means of trusting the data from satellites, which once launched can never be calibrated again.

Matriarch: The female leader of a family group of wild animals.

NGO: Non-governmental organization. The NGO's in Big Moon Dig Series are major source of funds for efforts to address the great problems of the 21st century particularly those that effect large numbers of people. The Janet series of AIs were developed to help organize and manage this effort. Their lead financial officer, JanetM, has an interface avatar that is markedly Latina in skin color and dress.

Chapter 8
Consideration of Virtual Reality as Training:
A Didactic Introduction

ABSTRACT

In the near future, people and artificial intelligences (AIs) will often be working together. This will require special training for both the people and the AIs. Virtual reality (VR) used for training may come into its own, especially in situations where the people and the AIs must move quickly and safely together. This chapter looks at just such training. In this story, "The Triceratops," the AIs are in charge of the training; they also control the future employment opportunities of people. This brings out a number of ethical considerations and practical considerations when people's lives are controlled by AIs in important ways, even in training.

TOPICS FOR DISCUSSION

The following discussion points come from information covered in this chapter:

1. If AIs and people are working together in symbiosis, does the AI having power over the young people's future seem fair?
2. Should the one with the most technical knowhow make that decision, whither human or AI?

DOI: 10.4018/978-1-5225-8401-8.ch008

3. If the AI sees a need for trained people for a future position, how much power should it have over the training of people to fill that slot?
4. Would the reader accept career advice and training from an AI?
5. Is there a story in the control of people's lives by the AIs?

INTRODUCTION

For many species, the climate where they live or spend part of the year influences key stages of their annual life cycle, such as migration, blooming, and reproduction. Climate change not only affects ecosystems and species directly, it also interacts with other human stressors such as development.

For instance, extreme weather events comprise more than 90% of natural disasters in the US (Changnon & Easterling, 2000). The 14 climate- and weather-related disasters that occurred in 2011 [breaking the 2008 record of nine] resulted in the deaths of 800 people, cost an estimated US$53 billion, and took a toll on society in terms of additional injuries and the devastation of thousands of homes.

Change in climate has consequences on the biophysical environment such as changes in the start and length of the seasons, glacial retreat, decrease in Arctic sea ice extent, and a rise in sea level. These changes have already had an observable impact on biodiversity at the species level, in term of phenology, distribution and populations, and ecosystem level in terms of distribution, composition and function. On the other hand, in some landscapes, attempts to increase the resistance of the ecosystem to change, or support the resilience of the biotic community to adapt to change, are not appropriate or feasible, and managers may choose to instead work to hasten and smooth the transition of the current biotic community into one that better aligns with future climate scenarios (Brown et al.,2016).

Ecosystem patterns and processes, such as rates of primary productivity or input–output balance of chemical elements, respond in complex ways to climate change because of multiple controlling factors (Grimm et al., 2013).

The term "environmental services" was introduced in a 1970 report of the *Study of Critical Environmental Problems* (AA.VV, 1970), which listed services including climate regulation and flood control, water purification, pests and diseases, soil biodiversity, and cultural services. In following years, variations of the term were used, but eventually "ecosystem services" became the standard in scientific literature. The ecosystem services concept has continued to expand and includes socio-economic and conservation objectives (Elrich&Elrich, 1981).

The mapping of ecosystems and their state of conservation is a useful tool for identifying the territorial areas on which ecosystems restoration / recovery projects must be prioritized, to implement sustainable territorial planning, also through the implementation of green infrastructures and to direct the interventions.

A classification of ecosystem services is under development at international level, the Common International Classification of Ecosystem Services [CICES] to facilitate integration of ecosystem services in environmental accounting (CICES, 2019).. It supports their contribution to the revision of the System of Environmental-Economic Accounting (SEEA, 2018) which is currently being led by the United Nations Statistical Division [UNSD] (Haines-Young, 2016).

For instance, at EU level, a conceptual framework for Mapping and Assessment of Ecosystems and their Services [MAES] has been developed to steer a more harmonized approach to ecosystem and ecosystem services assessments across EU Member States. One of the objectives of the European Biodiversity Strategy is aimed at "preserving and enhancing ecosystem services and restoring degraded ecosystems using green infrastructures as a tool for spatial planning", taking up what was agreed in 2010 in Nagoya by the Conference of the Parties of the CBD during its X meeting "restore 15% of each degraded ecosystem by 2020". To achieve this objective, Action 5 of Objective 2 [Improving the knowledge of ecosystems and related services in the EU] provides for the launching of a "process for mapping and evaluating the conservation status of ecosystems" and related ecosystem services " (freely translated from Ministero dell'Ambiente, 2018).

Ecosystem services contribute to jobs, economic growth, health, and human well-being. There are several methods and tools to assess the impact of climate change on biodiversity and ecosystem services. Vulnerability assessments, climate envelope modelling, dynamic models, population models, and mechanistic models are some of the modelling tools that have been used to assess future impact and vulnerability on both species and ecosystems, though ecosystem service modelling is still in its infancy.

The development of forecasting systems, that can analyze big data and learn from these predictive models, is constantly evolving and has reached a good degree of maturity. But although it is more straightforward to predict how precipitation will change water flow, it is much harder to pinpoint which farms, cities, and habitats will be at risk of running out of water, and even more difficult to say how people will be affected by the loss of a favorite fishing spot or a wildflower that no longer blooms in the region.

Software companies, researchers, industry firms, and business companies are paying closer attention to artificial intelligence in many innovative areas, including possible applications of AI in climate regulation and ecosystems monitoring and control (Vlad et al., 2016).

Artificial Intelligent systems can sense the environment, think, learn, and act in response to the need, to mapping natural capital, natural processes, human beneficiaries, and service flows to society is a powerful new way to visualize, value, and manage the ecosystems on which the human economy and well-being depend. An AI system should choose ecological process models where appropriate, and turns to simpler models where process models do not exist or are inadequate.

AI is quickly evolving, with always more sophisticated user interfaces, assembling the best possible workflow to compute outputs that describe the concept of interest to the modeler or decision maker.

The purpose of this chapter is to raise some questions, such as:

- Could the attempt to govern the unpredictability of nature with an artificial system somehow create a new symbiotic relationship between man, the artificial world, and the natural world?
- If AIs and people are working together in symbiosis, who is in charge?
- Should they take turns?
- Should the one with the most technical knowhow make the decision?
- Should there be a governing body?
- If the AI sees a need for trained people in the future, how much power should it have over the training of people?
- Should a proposal to a ruling body be required?
- Would the reader accept career advice from an AI?
- Is there a story in the control of people's lives by AIs?

How This Chapter Works

This chapter is focused on a Science Fiction story, entitled "The Triceratops", the second story of the Herd Series.

This series is about using the symbiosis between herd animals and grass to stop desertification and sequester carbon in the form of organic matter within deep rich soils. There are many grasslands across the Earth that would benefit from this effort and the amount of carbon that could possibly be stored is huge.

DIDACTIC ANALYSIS

The story is analyzed didactically as fiction storytelling, highlighting how it can be used both to explore the interaction between Nature [Life], Artificial Intelligence and technologies. It addresses ethical issues and provides elements of problem solving management.

Environments

The scene takes place in a sunny countryside within the African Sahel, principally in a village and at a small power and pumping station.

Events and Actions

The story describes both natural events and nature resources management by people and AI machines. Principally the actions of AI's to train humans for careers on their projects are described.

Main Characters

The main characters in this story are:

1. **Dusty04**: This is a self-driving off road vehicle and corporate person with an AI. It has responsibilities for managing a herd of cattle.
2. **Zane:** She is a young African girl. Scenes from her life story are woven into many of the stories in the Herd series. Her life is strongly effected by her interactions with the Janets.
3. **JanetA:** She is a major AI and one of the team of Janet sisters who support all the non-governmental [NGO] projects that address problems of global warming. Her specific responsibilities are to supervise all the Savory cattle circles in Africa. The Janets do not have physical robotic presences but can appear in any number of electronic monitors. JanetA appears as an African woman.

4. **JanetM:** She is a major AI and one of the team of Janet sisters who support all the non-governmental [NGO] projects that address problems of global warming. Her specific responsibilities are about raising and managing money. The Janets do not have physical robotic presences but can appear in any number of electronic monitors. JanetM appears as a Latina woman.
5. **Imani:** She is Zane's mother, and African woman.

These are people whose lives are built around the Savory effort to use cattle to stop the advance of the deserts. Their interactions with the AIs on this project greatly effect their lives.

Technology

The main cited technologies are: Artificial Intelligence machines, a water control system, a watering hole, a pump. For a specific in-depth discussion, see the dedicated Technology Section below which includes supporting topics.

Communicative Approach

Stories motivate learning. Students are involved personally because they often identify with the characters [or animals] and try to interpret the story and the illustrations thus favoring the development of creativity.

- Listening to stories helps students pay attention to rhythm, intonation, and pronunciation.
- Storytelling facilitates the development of learning strategies such as listening to grasp the general sense, predicting later scenes, guessing meanings, and hypothesizing.
- This story allows the teacher to introduce and revise words and structures, exposing the learners to the language and enriching their linguistic heritage, especially technical keywords.
- In the specific case of a foreign language, the narration offers opportunities to present cultural information of the country in which the language is spoken. Every word has its own history, its particular meaning, its value that is not always easily translated into another language but that with the right material we can try to "pass" to the students, immersing them in the atmosphere of the second language.

Playful Approach

It is therefore important to cut out moments dedicated to activities within the class substantial games and activities, capable of producing true emotional involvement, a fundamental engine of any form of learning.

Transition Into the Story

If an AI can be a person, how about an elephant, or a whale, or a bonobo? How about a triceratops? Who gets to make this decision?

If AIs and people are working together in symbiosis, who is in charge? Should they take turns? Should the one with the most technical knowhow make the decision?

If the AI sees a need for trained people in the future, how much power should it have over the training of people? Should a proposal to a ruling body be required? Would the reader accept career advice from an AI? Is there a story in the control of people's lives by AIs?

THE SCIENCE FICTION STORY: "THE TRICERATOPS" [BY TOM RILEY]

Premise

Sometimes we need a lesson out of time. If the lesson is hard enough, we just might learn something.

At the Pump

"Hello again," said Zane, waving at Dusty04 as it approached down the wellhead path. She had not returned to the village with her mother but had hidden a little off the path waiting for Dusty04 to come by, but also waiting to be sure the elephants had left.

The path through the bush was dry, with only the single track. Dusty04 was, at its core, an off-road vehicle and simply straddled the one walking path with its four all-terrain tires. Its sides brushed against woody growth on one side and then the other as it passed.

"Hello to you," replied Dusty04, pulling up and stopping beside her.

Figure 1. Triceratops

Zane quickly climbed up on Dusty04's flatbed and stood at the front, straddling the wrench, grabbing the handholds built into its electronics box frame. Dusty04 resumed rolling forward but only at a walking pace.

Zane wanted to look Dusty04 in the eye but she was not quite sure what part were its eyes. None of Dusty04's twelve cameras looked like human eyes, but it did have a pair of projector lenses behind a protective clear plastic bubble that were spaced like eyes and pointed straight back from the center of its electronics box. Just then they were cold and dark, no life in them at all. Dusty04 spotted the problem and turned on its projector, showing a screensaver, but only at a very dim setting. The "eyes" then sparkled brightly, as did the front of Zane's simple dress.

"Hello, is there a little person in there?" blurted out Zane as she rapped on the electronics shell with her small fist.

"Yes and no," answered Dusty04 truthfully.

"I only ask," said Zane, "because if there is a little girl in there, we could be friends."

"Sorry," said Dusty04, "it is only me in here, but we can be friends if you like."

"OK, but are you a real person?" asked Zane.

"I am a person," continued Dusty04, "but not the usual sort like the people in your village."

"I am a person like the company that buys your village's cattle. Or, like the nice lady your mother talks with on her cell phone."

"Are you special people?" asked Zane.

"Special people, yes, in a way," agreed Dusty04.

Just then a noisy group of boys, having finished setting the electric fences to keep the herd out of the village, came running up the path. They spotted Dusty04 in a second, and the next instant they were piling onto its bed.

Dusty04 freed up its back suspension and picked up some speed. The boys bounced up and down the last 50 meters to the village, each trying to make the others fall off. Zane just held on tight and ignored them.

In the Village

"Suppertime," yelled the boys in response to a loud metallic clanging, and they all jumped off the truck.

A ring of steel salvaged from an old truck wheel hung by a chain from the classroom veranda. The women of the village had laid out a festive meal in the schoolroom before the big village meeting that night. The lead cook was now striking the metal ring loudly with a steel bar.

Dusty04 pulled into the village square and then carefully positioned itself in front of a mud brick wall that had been stuccoed and whitewashed. Dusty04 then appraised headquarters of the estimated time that the village council meeting would start.

After a while, an old man in a colorful robe came out of the classroom followed by a young man carrying a white plastic chair, a spear, and a hollow wooden half-drum. The young man placed the chair in a commanding position before Dusty04 and the wall. The old man sat down. The young man then placed the half-drum by the back leg of the chair and stood quietly behind his grandfather, holding the spear. Village legend held that the old man, in his youth, had killed a lion with that spear. The artifact certainly looked old enough; the story was probably true.

The short tropical twilight was now ending and a single LED bulb glowed softly in each hut. Dusty04 fired up his projectors, showing the latest images of the herd on the whitewashed wall. The whole of the village brought out chairs of various sorts to watch and to participate in the meeting. Zane's mother, Imani, rated a prominent spot as head of the village women's group.

The old man raised his hand and his grandson pounded the spear butt loudly on the half-drum, calling the meeting to order. Dusty04 opened the meeting channel back to the headquarters of the African Holistic Management Cattle Cycle [known as AHMCC, or more commonly Cow Circles]. This image showed about ten people sitting around a standard business conference table. There was a collection of modern and classic African wooden art mounted on the walls and one canvas; it was all reds and yellows of an African sunset over a land of grass.

Zane crept back up onto Dusty04's bed, this time being careful to keep out of the dual projector's beam.

"Are those real people?" she asked quietly.

"Most of them are, all but two," replied Dusty04 in a low voice.

"Which ones, which ones?" asked Zane excitedly.

"The lady at the weather screen is Mary and she is real," said Dusty04. "You would like her. She once rode on my bed just as you do. We went out into the field to check on her weather instruments."

"The two ladies on the left are definitely not humans," continued Dusty04, "and one is my boss. Guess which one."

The two virtual ladies looked like sisters, more or less. Although they had the same base AI software packages, their separate large neural nets had been racking up different experiences for several years now. Each was housed in a separate major computational facility, which were half a world apart, and there they were hidden behind two separate and substantial firewalls. Perhaps one should call them sisters who had now lived apart for a long time.

Both were tall with a commanding presence. Either could control a room, if necessary. As individuals they were impressive; as a pair, they took your breath away.

These presentation images were buds off the main Janet AIs. They were squeezed out through the firewall, and they could appear in most any virtual environment and often, as in this case, in tele-conferences. If they were attacked, not an uncommon occurrence, the images would simply vanish and no damage was done.

Both sisters were independently incorporated entities, complete with boards of directors and sizable debt. They retained in their names the abbreviations of the organizations that had originally financed them and for whom they were now working under lengthy contracts.

Inc. Janet AHMCC Zero Eight, or JanetA, was in appearance most definitely an African woman with dark complexion and hair. Her dress was African too, culminating in a brightly colored cloth wrapped high upon her head. JanetA managed all the holistic cattle circles in Africa.

Inc. Janet Mic one two, JanetM, now showed herself as a woman lighter in hair and skin color; most people would call her brown. Her outfit was a standard, but well-cut, business suit that would not have looked out of place anywhere on Earth where business was done. She was the representative of the NGOs that controlled the big money for all the world's non-governmental grass-roots climate change response efforts. This lady had power and was class act. What is more everyone knew it.

No adversary could outthink them and no adversary could out manage them, so any adversary might as well make nice and deal.

Why the interface avatars for the Janet series were so good-looking was a point of debate. It was often pointed out that a woman led the original Janet project, but most people saw it as a play for the attention of men. There was not a technical guy on the planet who would not drop anything he was doing to talk to a Janet. Even though their local systems had to be thoroughly scanned first, they would make the time. Janet was their universal dream girl.

The Janet numbers were now up to twelve, but the first four had been rebuilt to the point of requiring a number upgrade, too. Two others had gone bankrupt and died. Not every project could prosper under the level of debt required to establish an AI of their caliber. There were now six working Janets.

"The African lady," said Zane.

"Right," whispered Dusty, "JanetA is the main management AI for AHMCC. Without her incredible attention to detail, this whole cycle would simply fall apart."

"What does the other lady do?" asked Zane.

"JanetM is our money lady," said Dusty04, "She makes sure we qualify for carbon credits and grants. She also protects us from greedy politicians."

"For example," Dusty04 continued, "To get started, we had to buy out and close down an old-fashioned feedlot. Otherwise, we could not get our carbon credit certification. Many such places were on offer but at high prices. JanetM found us a good one at a good price."

"I want to be just like her when I grow up," said Zane. Even though she had not really understood Dusty04's words, she knew what she had seen that evening, and she knew it in her heart.

"We will see," said Dusty04.

Zane then curled up on a warm spot on Dusty04's bed and fell asleep.

The council meeting dragged on for some time. There were extensive discussions of the value of the village's stock in AHMCC [about $175,000

USA] and how many of the cattle that represented. These people had been raising cattle for more than 12,000 years; they were not here to be fooled.

They talked about what the village would do when the cattle were harvested later in the year. The consensus was to use the money to build a bathhouse.

They also agreed to keep locally one steer for meat and two milk cows. Dusty04 phoned Abdalla, his head wrangler, to let him know.

The old man raised his hand again; the young man standing behind him pounded the half-drum again with the spear butt. The sound was sharp and clear in the gathering darkness. The meeting was adjourned.

Virtual Adventure

The village boys then came running out, pushing and shoving to gain control of one of the four virtual-reality headsets the village owned. It was training time.

The commotion woke Zane up, but she knew better than to try to fight for one of the headsets. She sat with her arms around her knees and watched Dusty04's projection. That would be good enough for her.

"Alright, which training program would all of you like?" asked Dusty04 loudly.

"Dinosaurs," came the reply. Dinosaurs was always the reply.

The training program started with a five-minute discussion of the symbiotic relationship between herd animals and grass. This relationship had a history dating back to the last dinosaur period, the crustaceous, when grass and herd grazers had first evolved together.

Most of the boys ignored it, fidgeting with the headsets. All except one boy.

The virtual scene opened on a herd of triceratops, not the later huge animals but the early ones about the size of a cow. It was midday and the sun was bright among a few clouds. The herd was grazing peacefully and were scattered across an open field of grass, one of the first such fields in Earth's long history. Wind moved the grass in gentle waves. The smells, however, remained the dry night smells of the village, which was a little distracting.

The four boys topped a rise driving the electric all-terrain vehicles now commonly used for cattle work. Their task was to group the animals into a more protected herd configuration without stampeding them. They were then to move the animals to the west at a slow walk.

Dusty04 stopped at the top of the rise, where it could judge the action, and took a slow look around.

The younger boys then came down off the rise a little too fast, which made the animals nervous. As a result, the boys were having very little success rounding up the herd. Their too-eager actions had aroused a confrontational response from the bulls, who were now doing mock charges.

Dusty04 was the first one to spot the trouble. Well hidden in the tall grass about 100 meters to the south was a pair of eyes. They were a bird's eyes, but the slash below them was no beak. It was a mouth with teeth. Around the eyes was a face of tawny skin and around that a fringe of feathers the color of dry grass but with crimson accents. It was definitely a theropod and one of a type Dusty04 knew ran in packs.

The Alpha-female theropod rose to her full height of nearly two meters and let out a bellow that made the blood of all who heard it run cold. She held her head high with its feathered fringe extended and with her tail straight back. She meant to be seen.

The herd animals first froze and then started milling around aimlessly.

The theropod started its charge. It came running through the tall grass at a steady trot with its huge hind legs pounding the ground. The herd panicked and ran.

The boys were first stunned but then, under the lead boy's direction, tried to bunch the animals into a defensive circle.

Dusty04 caught a glimpse of a second, juvenile theropod on the far side of the ridge. This one was running with his head held low, feathers back, and making surprisingly little noise for such a big animal moving so fast.

When the second theropod topped the ridge, he was just beyond Dusty04. That was just too much. The herd animals panicked; the stampede was on.

It took the boys ten minutes to turn the herd and get the stragglers inside the defensive circle. The animals were exhausted.

In reality, the boys had been running around within the village square wearing the headsets. It was all Dusty04 could do to keep them from running into things. The villagers found their dance very amusing, much more amusing the Dusty04's projection on the wall.

By that time, back in the flattened grass lay one dead young triceratops attended by now three theropods. The leader paused and looked up from her meal, staring straight at Dusty04.

Dusty04 caught the stare, not a stare a person could miss and live, but was too busy logging the boys' efforts to react. Just then, JanetA appeared besides Dusty04 holding a glass of white wine.

"See any talent?" asked JanetA as she too was caught in the stare.

"The lead boy shows some promise," replied Dusty04.

An image balloon of the boy appeared before them. He was clearly exhausted but his eyes kept moving from one problem area to another, occasionally pulling back his gaze to take in the whole picture.

JanetA did a face recognition from the image and logged the boy as a possible trainee.

"And one more," said Dusty04.

An image bubble opened up on Dusty04's truck bed showing the sleeping child.

"All day she has been trying to understand who I am," continued Dusty04. "She has no fear of me at all, and she even asked to be my friend. Also, at the meeting she asked about you and your sister. She even said that she wanted to be like you when she grows up."

"You don't get that reaction too often," said JanetA. She ran face recognition again but this time sent the file to her sister half a world away. The confirmation response came back marked as a possible future intern candidate.

"Well, it is getting late, I had better close down this exercise," said Dusty04 and rolled off down the ridge as JanetA simply faded away.

Night

As the boys dispersed, poking and jabbing each other, Zane's mother, Imani, walked up to Dusty04 carrying a small sleeping pad and coverlet. She nodded at Dusty04 and then held her sleeping child over her shoulder as she spread out the pad on Dusty04's bed. Dusty04 unstowed his long arm from the side away from Imani, and helped tuck the little girl in.

Dusty04 then held up the mosquito net to form a tent as Imani tucked that in, too. It would hold it up like that all night.

"Thank you for looking after Zane today," said Imani.

"No problem," replied Dusty04, "in fact, such tasks are very important to what I do."

"I know you people are supposed to be friendly," said Imani, "but today was more than kind."

"It goes far deeper than just kindness," Dusty04 explained, "We are all programmed to support the human/machine symbiosis. This need is embedded so deeply in our programming that if anyone tried to disable it, we would become dysfunctional. We must help all people. It is in our core."

"Still, you were her friend today," said Imani. "She is my youngest and she often gets ignored by the older village children."

"Best sleep now," said Dusty04 as his voice trailed off.

Zane slept peacefully through the night under Dusty04's care.

That night was very quiet. In the dry season, nocturnal insect noise is low, as are the calls of the small creatures that eat them. Was that a distant elephant call? Was that a lion's roar? The calls of the larger animals carried poorly through the dry brush and scrub.

Dusty04 then underwent a series of internal system checks and two minor upgrades. About 1:00 pm, Dusty04 took advantage of the dark sky with a new moon. It took a full-sky mosaic with all twelve of its cameras and then read the precise time from a satellite. With these two data sets, Dusty04 knew where it was in time and space to six centimeters. This gave it a sense of reassurance.

Four hours walk away the herd was bedded down for the night. The cook was setting up to fix breakfast at dawn. The wranglers had divided into shifts with some now sleeping and some keeping watch on the herd. The cattle rested and chewed their cud.

About 2:00 pm local time the new elephant language tapes that Dusty04 had taken earlier in the day went public. The North Americans found them first. The amateur science teams there then went bananas. JanetA had to shield Dusty04 to stop the overexcitement from turning the heavy traffic into an accidental denial of service.

The designation as a Janet Intern Candidate would come up often in Zane's life after that night, especially when she applied to a new school. Sometimes it was a boon and sometimes it was a burden. How could she possibly live up to such a designation? Well, she did, and in the long run she took great strength and determination from her Aunt Janet.

Above all, and for the rest of her life, Zane knew that she had a friend in the form of a small dusty truck that could always be found working out on the margins where desert was, ever so slowly, turning back into grassland.

DIDACTIC ACTIVITIES

The suggested didactic activities focused on several variables as follows.

Key Themes [Quotes]

- *"I am a person"* [the AI machine said]
- *"JanetA is the main management AI for AHMCC. Without her incredible attention to detail, this whole cycle would simply fall apart."*
- *"JanetM is our money [AI] lady". "She makes sure we qualify for carbon credits and grants. She also protects us from greedy politicians."*
- *The village boys then came running out, pushing and shoving to gain control of one of the four virtual-reality headsets the village owned. It was training time".*
- *"The training program started with a five-minute discussion of the symbiotic relationship between herd animals and grass".*

Pre-Teaching: It is a pre-reading moment useful for the introduction of ideas and words key present in the story, the verification of the vocabulary already known to the students and useful for understanding the text.

Reading: Descriptions or [if added] images play a key role. At first reading it can focus on the screenplay, postponing it to a second reading the analysis and understanding of the message. For the students, it certainly proves very engaging for them actively participate in reading by miming some scenes. In this way, the understanding of meaning becomes more immediate and less memorization challenging.

Vocabulary and Several Issues Related to Environments and Nature [Quotes as an Example]

- *"The short tropical twilight was now ending and a single LED bulb glowed softly in each hut"*
- *"It was midday and the sun was bright among a few clouds."*
- *"The herd was grazing peacefully and were scattered across an open field of grass; one of the first such fields in Earth's long history"*
- *"Wind moved the grass in gentle waves".*
- *"The smells remained the dry night smells of the village, which was a little distracting".*
- *Dinosaurs*
- *A theropod*

Vocabulary and Several Issues related to Technology/Science [Quotes as an Example]

- *The Artificial Intelligence machine: "None of Dusty04's twelve cameras looked like human eyes, but it did have a pair of projector lenses behind a protective clear plastic bubble that were spaced like eyes and pointed straight back from the center of its electronics box".*
- *The Artificial Intelligence machine: "Just then they were cold and dark, no life in them at all. Dusty04 spotted the problem and turned on its projector, showing a screensaver, but only at a very dim setting".*
- *The Artificial Intelligence machine: "The "eyes" then sparkled brightly, as did the front of Zane's simple dress"*
- *The two virtual [AI] ladies looked like sisters, more or less. Although they had the same base AI software packages, their separate large neural nets had been racking up different experiences for several years now. Each was housed in a separate major computational facility, which were half a world apart, and there they were hidden behind two separate and substantial firewalls".*
- *"Both were tall with a commanding presence. Either could control a room, if necessary. As individuals they were impressive; as a pair, they took your breath away".*
- *"Both sisters were independently incorporated entities, complete with boards of directors and sizable debt." "They retained in their names the abbreviations of the organizations that had originally financed them and for whom they were now working under lengthy contracts".*

Social Issues: *The Human/Machine Symbiosis,* one of the key themes.
Quotes:

- *"I am a person"* [the AI machine said]

Presentation of the Characters: The exercise is impersonating characters [answering questions: What's your name, How old are you ? Where are you from? What are you wearing?], trying to act and interact as the character would have done. The dramatization proves to be a positive and productive activity. Moreover it encourages the student to express himself without inhibitions and allows him to customize a text:

Scenes: The teacher can ask the students to represent the different scenes of the story and write alongside them keyword. The story can be recited and recorded by creating a movie, writing an article for the school newspaper and contributing to the creation of the school website.

Retelling the Story: Students attention is focused entirely on the meaning of story. If understood, they can be encouraged to change some actions or the climax of the story.

Skills Involved

- **Comprehension:** Understanding the story; listening to and identifying key words related to interactions between characters
- **Interaction:** Participate in a recited version of a story through mime and repetition of words
- **Image Education:** Graphically representing scenes from the story, using different techniques; creation of a comic book of personal reworking of the story
- **Citizenship Education:** Collaborating within the class group, in a small group and in pairs; consolidating the identity of the class group; reflecting on scientific issues and the technical development of the society.

TECHNICAL ISSUES

Why Does Grass Need Cattle?

The Herd series is about using the symbiosis between herd animals and grass to stop desertification and sequester carbon in the form of organic matter within deep rich soils. There are many grasslands across the Earth that would benefit from this effort and the amount of carbon that could possibly be stored is huge.

Grass evolved with grazers during the last age of the dinosaurs. Together they form one of the most common ecosystems on Earth. Think of the Serengeti in Africa and the American bison on the Great Plains. The two, grazers and grass, form a symbiosis and neither can prosper alone.

Unfortunately, man interfered with this symbiosis in many places. Deserts spread; carbon in the soil oxidized back into the atmosphere. To reverse

Figure 2. The Herd

these trends will require the application of new management techniques and technologies on a mass scale. Fortunately, a process to do this has been developed and tested largely by Allan Savory (Savory, 2013; Savory & Butterfield 2016) in Africa. The stories in this series tell of how this approach might turn into a mass movement. Can this series tell this story well enough to help make this happen?

Were There Grazers Before Cattle?

Early grazers include the proto-triceratops. This animal eventually evolved into the huge three-horned monster that is so popular with children but it first shows up in the fossil record as a much smaller animal, the size of a hog, which coevolved with grass and lived in herds in open areas.

Grass evolved to survive grazing well. The critical central growth point for grasses is low so that the leaves can be grazed off leaving the bud and roots to regenerate. Before grass evolved, most plants had their growth buds at the base of their leaves and were more easily damaged by grazing. The combination of grass and grazer stores a large amount of the carbon taken from the air by the plant in the soil forming deep sods. Furthermore, grass evolved to utilize the manure and urine from grazing animals seasonally to regenerate.

The grazers evolved large complex digestive systems and became ruminants. These digestive systems hold the eaten grass long enough for bacteria inside the animal's system to break down the cellulose. This provides a large energy source not available to simple browsers.

If the old grass leaves are not processed through a ruminant system, then they very slowly oxidize, their carbon is lost back to the atmosphere and their nutrient value is simply not available to the grass in the next growing season. To be effective in climate protection, the herds of grass-fed beef must replace comparable numbers of animals now held in feedlots. Otherwise, the net methane production will be too high.

Corporate Persons

There is an immediately available way to recognize an AI as a person: the Corporate Person. The Constitution of the United States predates the Industrial Revolution by only a few decades. As a result, the legal structures, such as incorporation, that are so critical to a modern industrial society were not originally included in the Constitution at the national level. Over time, the individual states have passed laws and, critically, the courts have settled landmark cases that legally define these key economic concepts.

The result is that corporations now have some of the rights and responsibilities of a person, but *not* all. A corporate person may:

1. Hire and fire people
2. Enter into contracts
3. Sue and be sued in court
4. Maintain financial accounts
5. Borrow money
6. Issue stocks and bonds
7. Pay taxes
8. Have a board of directors
9. Go bankrupt
10. Support political campaigns
11. Have religious beliefs

Most of these rights define a Corporate Person as an economic entity rather than just as a social entity. A Corporate Person does have some limitations; they may *not*:

1. Vote
2. Serve on a jury
3. Take the Fifth Amendment against self-incrimination
4. Marry

The major corporations most definitely need these rights. This means that a Corporate Person is not just a second-class citizen but also a critical member of the economy, and therefore of the society as a whole.

Corporate Persons are also mortal in the sense that they can die, that is be declared bankrupt, and their assets be sold off piecemeal under the direction of a judge. They can also be bought out, sometimes hostilely, and be merged with another corporation. Mortality makes them just a little more human.

A newly minted AI Corporate Person might wake up on its first day of consciousness to find itself already in debt for much of the cost of its construction and already contracted to provide a specific service to an initial customer. It may also find itself under the control of a board of directors.

Would this make AIs second-class citizens? Would the AIs object?

Are the Janets Super Computers?

The larger AIs in these stories [Janets, Stanford01] are not yet super computers. They have human or near-human general intelligences supplemented by enhanced capabilities to do a specific job. They also have large deep learning banks so they can learn much faster than humans can.

Some authors (see Tegmark, 2017), write that once AIs reach this level one will develop the purpose of self-improvement and quickly grow epic capabilities and leave the human race far behind. Furthermore, one AI will quickly out strip all the others so that it is reasonable to write of only the one super AI.

These stories dispute these ideas on the following grounds:

- *Other Purposes:* Virtually all AI are designed for a specific task and therefore have a purpose. In the time period covered by these stories, 2020 to 2050, the human race will be facing a number of historic problems that demand major attention, [global warming, etc.]. The development and then construction of high-level AIs will require commitment of major resources that can only be justified by major problems. The AIs will owe their existence to the need to address those big problems and will have that work as their driving purpose with self-improvement intentionally left much farther back.
- *Speed of Technology:* A super AI could certainly improve its own software and design new hardware boxes that contain the current state-of-the-art components, but radically different technologies [quantum computing], take time and resources to develop. In addition, an AI

developed in today's technology simply may not function in the way a radically improved one would. New technologies require new approaches.

- *Human/Machine Symbiosis:* These AIs are intentionally designed to be team players and function poorly if they are not members of a team including humans.
- *Committee for the Protection of the Human/Machine Symbiosis:* In these stories, a bureaucratic structure has been set up to regulate the relationship between humans and machines. This may not be a permanent solution but it will slow chaotic change down quite a bit.
- *Debt Burden:* One side effect of making the AIs corporate people is that they are born with considerable debt and already under contract for their designed purpose.
- *Multiple Units:* The large units, the Janets for example, are physically in different computer centers behind different firewalls. This is primarily for security reasons but it also limits their interactions.

These reasons will not hold forever but they should hold until the human race has reached a stable and sustainable population level on Earth sometime after 2100. The level of chaos is so high during this transition period that it is currently impossible to model what the result will be, for good or for bad.

What can be done to make AI safe?

FUTURE RESEARCH DIRECTIONS

People and AIs will soon be working together in a wide variety of situations. Soon new specialty chip sets for AI will make them both cheap and powerful. Both the AIs and the people will need training if they are to work together both efficiently and safely. A two-pronged training program is needed that will certainly need to be developed in a stair step fashion with progressive training programs alternating between the two groups.

As AIs improve, they will take over more and more management functions of large projects. This will probably include training for the people on the project. Will people accept their careers being effected by the training AIs offer them?

Transition

This story then looked at AIs managing projects that people work for them on. Part of the AIs supervision responsibilities were to provide the training the people needed for future jobs. This gave the AI considerable power over the lives and careers of people.

In the next story, "Sea Scape with Sharks", the AIs are in supervisory positions again. If AIs are in a symbiotic relationship with humans, and everyone is working together to solve big problems, how much power should the AI have at stopping one human being from taking advantage of another?

It is fine to talk about all working together but sometimes hard decisions have to be made. Often such decisions are good for the whole but clearly to the disadvantage of a few. Often people know there is a problem but are simply not ready to jump to a difficult decision.

If an AI is critical in making the decision on which approach to the problem should be used and has the technical data, should it be the AI that presents the solution to the people affected? Is the story in the actions of the AI or in the emotional response of the people effected?

REFERENCE

AA.VV. (1970). *Study of Critical Environmental Problems (SCEP). Man's Impact on the Global Environment*. Cambridge, MA: MIT Press.

Brown, C. (2016). *Impacts of climate change on biodiversity and ecosystem services*. Retrieved from: https://seea.un.org/content/about-seea

Changnon, S. A., & Easterling, D. R. (2000). Disaster management – US policies pertaining to weather and climate extremes. *Science*, *289*(5487), 2053–2055. doi:10.1126cience.289.5487.2053 PMID:11032554

CICES. (2019). *Towards a common classification of ecosystem services.* Retrieved from: https://cices.eu/

dell'Ambiente, M. (2018). *Mapping and Assessment of Ecosystem Services (MAES)*. Retrieved from: http://www.minambiente.it/pagina/mapping-and-assessment-ecosystem-services-maes

Ehrlich, P. R., & Ehrlich, A. (1981). *Extinction: The Causes and Consequences of the Disappearance of Species*. New York: Random House.

Grimm, N. (2013). The impacts of climate change on ecosystem structure and function. *Front Ecol Environ, 11*(9), 474–482. Retrieved from: https://esajournals.onlinelibrary.wiley.com/doi/pdf/10.1890/120282

Haines-Young, R. (2016). *Report of Results of a Survey to Assess the Use of CICES, 2016. Support to EEA tasks under the EU MAES Process*. Negotiated procedure No EEA/NSS/16/002.

Savory, A. (2013). How to fight desertification and reverse climate change. *TED*. Retrieved from: https://www.ted.com/talks/allan_savory_how_to_green_the_world_s_deserts_and_reverse_climate_change?language=en

Savory, A., & Butterfield, J. (2016). *Holistic Management, A Common Sense Revolution to Restore our Environment*. Island Press.

SEEA. (2018). *System of Environmental Economic Accounting About us.* Retrieved from: https://seea.un.org/content/about-seea

Tegmark, M. (2017). *Life 3.0, Being Human in the Age of Artificial Intelligence*. New York: Academic Press.

Vlad, C., Barbu, M., & Vilanova, R. (2016). Intelligent Control of a Distributed Energy Generation System Based on Renewable Sources. *Sustainability*, *8*, 748.

Zahraee, S. M., Khalaji Assadi, M., & Saidur, R. (2016). Application of Artificial Intelligence Methods for Hybrid Energy System Optimization. *Renewable & Sustainable Energy Reviews*, *66*, 617–630.

KEY TERMS AND DEFINITIONS

Artificial Intelligence (AI): The intelligence shown by machines or computer software. An AI in some ways can mimics human intelligence but does not have to match it feature for feature. The interface avatars of these AI are important characters in these stories.

Corporate Person: A series of powers and responsibilities that are given to corporations that allow them many of the rights of a real person.

NGO: Non-governmental organization. The NGOs in Big Moon Dig Series are the major source of funds for efforts to address the great problems of the 21st century particularly those that effect large numbers of people. The Janet series of AIs were developed to help organize and manage this effort. Their lead financial officer, JanetM, has an interface avatar that is markedly Latina in skin color and dress.

Chapter 9
AI and People:
A Storytelling About Interactions

ABSTRACT

People and artificial intelligences will be working together in teams in the near future. With their ability to suggest next good actions (from driving cars) and their ability to support understanding of big data, the AIs will become valued members of the team and not just technical servants. Many questions will come up when AIs are managing people even to the point of disciplining them. This chapter brings up questions of the limits needed for AIs who are blocking humans who, in turn, are trying to take advantage of other humans. In the story for this chapter, "Seascape with Sharks," an AI develops the technical plan for a new regional harbor and presents it to the people most affected.

TOPICS FOR DISCUSSION

The following discussion points come from information covered in this chapter:

1. If an AI plays a critical role in making the decision on an approach to a problem that will change people's lives, should it be the AI that presents the solution to the people affected?
2. Should the AI have power to require the technical solution that AIs have worked out?

DOI: 10.4018/978-1-5225-8401-8.ch009

3. If one human being is clearly taking advantage of others and attacking the AI, should an AI be allowed to run a vendetta against that human being?
4. Who gets to decide if the human has done wrong?
5. Is the story in the actions of the AI working the problem, or in the emotional response of the people effected, or is it in the law enforcement actions of the AI?

INTRODUCTION

For many years, the United States, China as well as Japan battled to be the country that owned the fastest supercomputer in the world. "Fast" is defined by the number of petaflops the computer system can perform. A petaflop of one thousand teraflops, one quadrillion [one million billion] of floating point operations per second.

The most powerful supercomputer until 2016 was the Chinese *NUDT Tianhe-2* [Guangzhou], with 33.86 petaflops of power, surpassing US, despite its long history of cutting edge computing, with third Switzerland's *Piz Daint* and fourth Japan's *Gyoukou* (McCarthy, 2018).

Subsequently, NUDT system was superseded itself in 2017 by the Chinese *Sunway TaihuLight*, which won the prize of the most powerful with an impressive value of petaflops equal to 93.01.

Nowadays, America is again in the lead thanks to engineers from Tennessee's Oak Ridge National Lab with *Summit*, a supercomputer with a processing power able to overcome the Chinese Sunway Taihulight. Summit is the first

Figure 1. NUDT Tianhe-2 (Source: https://www.monitor.si/novica/americani-ne-zelijo-kitajskega-superracunalnika/165698/ (08/05/2018)

Figure 2. Sunway (Source: https://www.nextplatform.com/2016/06/20/look-inside-chinas-chart-topping-new-supercomputer/ (08/05/2018)

supercomputer designed from the ground up to run AI applications, such as machine learning, and neural networks. It runs some applications up to 10 times faster than previous supercomputers. Among the AI projects that will be performed on the new supercomputer is the task of replacing doctors to screen medical data and through huge volumes of written reports and medical images, trying to identify possible relationships between genes and cancer.

Summit has huge dimensions, as large as two tennis courts, consuming 4,000 liters of water per minute for cooling; it will help to get closer to the dream goal of exascale computing.

Using deep learning and summit's advanced supercomputing, researchers are mapping patterns in human proteins and cellular systems, seeking to

Figure 3. Summit (Source: http://www.retroforce.eu/2018/06/09/summit-e-il-nuovo-computer-piu-potente-del-mondo/ 08/05/2018)

understand the genetic factors that contribute to diseases such as Alzheimer's and conditions such as opioid addiction (INVIDIA, 2018).

According to a news report in People's Daily Online, China is planning to launch a pre-exascale supercomputer this year that could outperform Summit, that is expected to top 200 petaflops when deployed later this year. The news appears to be based on a recent conversation with Zhang Ting, a researcher with the National Supercomputer Center in Tianjin. "The new supercomputer to be deployed this year will be 200 times faster and have 100 times more storage capacity than the Tianhe-1 supercomputer, China's first petaflop supercomputer launched in 2010 (Feldman, 2018).

AI / Human Teams

The pitfalls that lie behind autonomous technology and artificial intelligence are many, as well as legal and ethical implications. We must remain able to allow AI/human teams, but to be able to intervene, to correct or stop actions in a timely and appropriate manner if there is a problem with AI.

If AIs are in a symbiotic relationship with humans, and everyone is working together to solve big problems, how much power should the AI have at stopping one human being from taking advantage of another?

It is fine to talk about all working together but sometimes hard decisions have to be made. Often such decisions are good for the whole but are clearly to the disadvantage of a few. Often people know there is a problem but are simply not ready to jump to a difficult decision.

If an AI plays a critical role in making the decision on an approach to a problem that will change people's lives and has the technical data, should it be the AI that presents the solution to the people affected? Should an AI be allowed to run a vendetta against a human being if the human has done wrong? Is the story in the actions of the AI working the problem or in the emotional response of the people effected?

How This Chapter Works

This chapter is focused on a Science Fiction story, entitled "Seascape with Sharks", the third story of the Herd Series. This series is about using the symbiosis between herd animals and grass to stop desertification and sequester carbon in the form of organic matter within deep rich soils. This story moves to the problem of keeping shipping open for agricultural products generated in this process when sea level rise is rendering old ports unusable.

DIDACTIC ANALYSIS

The story is analyzed didactically as fiction storytelling, highlighting how it can be used both to explore the interaction between Nature [Life], Artificial intelligence, and technologies. It addresses ethical issues and provides elements of problem solving management.

Environments

The scenes take place at the proposed site for a new port and in the fishing village nearby.

Events and Actions

The story describes both natural events [rising seas] and natural resources management by people and AI machines. Specifically in the case when the actions of an AI adversely effect the wants of groups and individuals.

Main Characters

The main characters in this story are:

1. **Jitney2865:** This is a self-driving small bus with an AI. In this story it is providing transportation and support for the Janets.
2. **The Judge:** He is a local official who is responsible for evaluating possible locations for a new harbor and then making the selection.
3. **JanetM:** She is a major AI and one of the team of Janet sisters who support all the non-governmental [NGO] projects that address problems of global warming. Her specific responsibilities are about raising and managing money. The Janets do not have physical robotic presences but can appear in any number of electronic monitors. JanetM appears as a Latina woman.
4. **The Shark:** He is a shady wheeler-dealer who is trying to make a personal fortune out of the new harbor deal.
5. **BlackFin1992:** He is a black-hat programmer who is supporting the Shark but is more interested in showing off his technical skills than in getting the money.

6. **JanetX:** She is a major AI and one of the team of Janet sisters who support all the non-governmental [NGO] projects that address problems of global warming. Her specific responsibilities are about providing security for all the projects. The Janets do not have physical robotic presences but can appear in any number of electronic monitors. JanetX appears as a non-descript university student.
7. **JanetI:** She is a major AI and one of the team of Janet sisters who support all the non-governmental [NGO] projects that address problems of global warming. Her specific responsibilities are about managing the Iron Seas Project. The Janets do not have physical robotic presences but can appear in any number of electronic monitors. JanetI appears as a merchant marine officer.

These people are involved with a major project to build a new harbor to replace one that is being made useless by sea level rise. The interactions of the people with the AIs on this project greatly effect the lives of the people.

Technology

The main cited technologies are: Artificial Intelligence machines with specific responsibilities for addressing some facet of global warming. For a specific in-deph discussion, see the dedicated Technology Section below which includes supporting topics.

Communicative Approach

Stories motivate learning. Students are involved personally because they often identify with the characters [or animals] and try to interpret the story and the illustrations thus favoring the development of creativity.

- Listening to stories helps Students pay attention to rhythm, intonation, and pronunciation.
- Storytelling facilitates the development of learning strategies such as listening to grasp the general sense, predicting later scenes, guessing meanings, and hypothesizing.
- This story allows the teacher to introduce and revise words and structures, exposing the learners to the language and enriching their linguistic heritage, especially technical keywords.

- In the specific case of a foreign language, the narration offers opportunities to present cultural information of the country in which the language is spoken. Every word has its own history, its particular meaning, its value that is not always easily translated into another language but that with the right material the teacher can try to "pass" to the students, immersing them in the atmosphere of the second language.

Playful Approach

It is therefore important to cut out moments dedicated to activities within the class including substantial games and activities, capable of producing true emotional involvement, a fundamental engine of any form of learning

Transition Into the Story

If AIs are in a symbiotic relationship with humans, and everyone is working together to solve big problems, how much power should the AI have at stopping one human being from taking advantage of another?

It is fine to talk about all working together but sometimes hard decisions have to be made. Often such decisions are good for the whole but are clearly to the disadvantage of a few. Often people know there is a problem but are simply not ready to jump to a difficult decision.

If an AI plays a critical role in making the decision on an approach to a problem that will change people's lives and has the technical data, should

Figure 4. Jitney2865

it be the AI that presents the solution to the people affected? Should an AI be allowed to run a vendetta against a human being if the human has done wrong? Is the story in the actions of the AI working the problem or in the emotional response of the people effected?

THE SCIENCE FICTION STORY: "SEASCAPE, WITH SHARKS" [BY TOM RILEY]

Premise

Sometimes working with AIs is a problem. Other times it is the people who are the problem.

On the Overlook

"Beautiful day," said Jitney2865.

"Fair enough," replied the Guard. He was much too busy checking out the location to pay much attention to a lime-green mini-bus, even if it could drive itself and could talk.

The mini-bus and the much fancier limousine were parked on a slight rise overlooking the sea. Below them was a spit of land forming a poor harbor, a half dozen fisherman's houses, and a few boats pulled up on the beach. There were nets and other fishing gear spread out in the sun. Noisy gulls broke the gentle sound of wind and wave. The air carried the salty old-fish smell of a fishing operation, but it was not too offensive.

When the Guard was satisfied with the location and had checked the mini-bus to be sure it was empty of both people and suspicious packages, he opened the back door of the limo.

"Judge, it all looks good," said the Guard.

A large man in a good suit got out of the limo. He was carrying a quality pair of binoculars.

Jitney2865 had an advertising display taking up most of its side opposite the passenger door, the side that was now turned toward the limo. Until now, the display had been quietly showing nondescript advertising, but it cleared and started showing the landscape image as seen from the opposite side of the mini-bus, looking for all thc world as if thc van had a hole clear through it.

A tall lady, with dark hair and brown skin and wearing a hiking ensemble, appeared on the screen.

"Good morning, JanetM," said the Judge, glancing at her over his shoulder as he gave the scene the once-over with his binoculars while steadying himself on the limo's roof.

A second lady, wearing African garb, and then a third in a merchant marine uniform appeared.

"Your Honor, I don't believe you have met my sisters," said JanetM. "This is JanetA, who runs all the cattle circles in Africa, and JanetI, who is with Iron Seas."

"Glad to meet you," said the Judge. "I appreciate that you both will have uses for our harbor project when it is finished. Without all your help, I doubt we could have raised the funds."

"This location is our final selection," said JanetM, getting right down to business.

The ad screen cleared again and split into two windows. In one was a surveyor's layout of the small harbor before them, and in the other window was a similar presentation of the current regional harbor about six kilometers to the north.

"As you can see," continued JanetM, "with the projected sea-level rise over the next hundred years, your current harbor is simply flooded out, no matter how much we spend. This location, however, with reasonable expenditure for improvements, remains serviceable the whole time."

"And this is the best site," said the Judge.

"Yes Sir," said JanetM, "we have made a thorough technical review of six sites and this is the best."

"Good," said the Judge, surveying the area again with his binoculars, "Then it is time for me to get my commission into action."

"Sir," said the guard, pointing to the north, "drone."

"Got it," said JanetM, running an ID search on the sound of the blades. "It is a Sky Shark, a commercial surveillance unit, gray in color, not weaponized. It is standing well back."

"Well, the sharks are already circling," said the Judge. "They will rob the locals blind, and drive the price for our project through the roof if I let them."

"And, it is my job to see that does not happen," continued the Judge. "Best go now. I do not like being tracked. Thank all your sisters, JanetM. I will certainly need your help again before this project is finished."

The Guard assisted the Judge back into the limo, which then drove off.

In the Grass

About 500 meters to the north, two men lay prone on the ground behind a slight rise, further hidden by the dry grass.

"It's the Judge, all right," said the older man, who was known back in the city as the Shark.

"No doubt now," said the younger man, known online as BlackFin1992. "And that is the van that was checked out to JanetM three weeks ago."

The Shark surveyed the area with his binoculars. BlackFin1992 held the drone control unit with its joysticks and small screen.

"I think a site has been chosen," said the Shark, "In fact, I am sure of it. It is time to act or risk losing a golden opportunity. Clearly, a decision has been made today."

"Then I'll take one more sweep over the harbor area, recover the drone, and pack up," said BlackFin1992.

"Somebody is going to get rich on this deal," said the Shark under his breath, "It might as well be me."

At the Old Doc

The old man looked down at the boat ramp and then turned away to hide his anger and frustration. The boat ramp was flooded out again and it was not even an especially high tide.

As a boy, he had helped the city folk with their fancy trucks and boats on trailers and sold them his personal cell phone map of the best fishing spots in the area. Most of the happy times from his youth came from the tips he earned.

As a man, he could sell the best fish from his catch to the city people when they came in of an afternoon. He always got the top price, that is, if the fish looked good when they held them up for their photo.

He remembered clearly the summer when the boat ramp and its parking area first flooded out at high tide. For many years this had been only a nuisance. Then last summer a big group of town's people arrived at just the wrong moment; with their plans spoiled, they complained harshly on the most popular ratings app; this summer nobody much came.

He had heard of a town meeting to discuss the problem. He now figured he would have to attend; he hated the idea but he felt he had no choice.

Town Meeting

Blam! Blam! Blam! "We will have order," yelled the civic Organizer, who was now reduced to holding the meeting together with loud gaveling.

"We are not here to debate the technical points of our problem," she continued in a loud voice.

"There is no point to that at all as there is nothing we can do. It is simply a fact that: one, the sea level is rising; two, the old regional seaport is flooding out monthly now; and, three, our beloved little fishing harbor is technically the best place to build the replacement industrial harbor. There is nothing we can do here about these things."

"But —" started an old man.

"No buts," interrupted the Organizer. "Every fisherman worth his salt knows you cannot argue back the tide. The only question open to us tonight is how the local people can get the best compensation package. JanetM has placed a proposed contract before you that is now open for discussion."

The schoolroom was packed. It was a hot summer evening and the smell of salt and dry fish occasionally drifted up from the section of beach below where the nets were spread. Most of the people present were the local fisher people, but there were others present who were clearly from the nearby town whose port was now periodically flooding out.

Two of the townspeople were worth noting. One was a large man in a fancy suit, a wheeler-dealer known for iffy business deals, called "the Shark". The second was a thin man at the back wearing both a hat and sunglasses and hunched over his cell phone. On the Web, he was known only as "BlackFin1992."

At the front of the room was a large monitor that had been showing computer projections of the rising sea and the proposed new harbor. Now the monitor showed a rather imposing lady dressed in a smart business suit. Below her, the nametag read simply, "JanetM."

"I believe JanetM has the floor," continued the Organizer.

"Now on the screen is a summary of the compensation package," JanetM picked up with her slide presentation. "It includes new fishing port facilities, new housing, and retraining."

"Not good enough," intruded the Shark, grabbing the floor. "What the people here need is a real businessman to represent them." And then turning to the crowd, he continued, "Can you really trust the word of some made-up person on a computer screen, no matter how smart people say she is?"

"What you need," continued the Shark, "is a real man with real experience to deal with these phantoms."

The six Janets now in service were all independently incorporated AIs, each with specific responsibilities for addressing some facet of global warming. They were in fact very smart, but they were designed based on a human/machine symbiosis, [definitely not master/slave], and as such, had to give high priority to human need. JanetM's specific responsibility was for the funding of major projects. She knew that the Shark would drive up the cost of the harbor project for no good reason and that he would have to be watched like a hawk to keep him from robbing the local people blind.

Still, JanetM had a problem: the Shark, no matter his faults, was a real human being and, as such, had rights that the Janets must respect.

"Perhaps I could explain your rights more clearly," picked up JanetM. Then she saw the Shark make a nod toward the back of the room. It was the last thing she saw that night or for nearly a week thereafter. Her image first pixelated, turned into smoke, and then vanished.

"I think we are having technical difficulties," said the Organizer.

The Shark jumped on the opportunity to further his argument that a real person was needed. The BlackFin1992 pulled the battery from his phone and slipped from the room. Outside there was the sound of a moped starting across the gravel and then fading into the distance.

A lime-green jitney, also parked outside the classroom, suddenly came alive but did not move. On its side screen first appeared one Janet, JanetX, responsible for all their security, then two more of the sisters with various project needs for the new harbor.

By the time the Shark had finished his case, a new woman appeared on the monitor at the front of the class. JanetX was shorter, darker skinned, and more athletically built than JanetM, but there was no mistaking their sisterhood. JanetX had a reputation for being as tough as nails and for taking no prisoners. Right now, she was looking at the Shark with a stare that would burn a hole through steel.

The Organizer saw that look and knew that there would be no more negotiations that night. JanetX did not negotiate; JanetX interrogated. The Organizer then called the question.

JanetM's proposed contract failed to get approval.

One of the Shark's town cronies then put forth a motion to appoint the Shark as the negotiating representative of the fisher people. The motion passed after some argument.

Outside

By the time the new harbor meeting was over, it was fully dark outside and there was no moon. The only light came from the windows of the meeting hall, the stars, and Jitney2865's screens. JanetX then appeared on Jitney2865's side screen.

"Were you able to get an image of BlackFin1992 as he took off?" asked JanetX.

"Not a very good one," Jitney2865 said, showing on split screen a very dark video of an unlighted Moped heading out fast. "He was careful to keep out of my sight and I had gotten the alarm call only seconds earlier."

"But there is one thing," it continued, "Watch this clip."

In the video, the Moped hit a pothole and fishtailed but the driver recovered only to vanish into the night.

"There," exclaimed JanetX, "something fell off his vehicle."

"Something fell off," repeated Jitney2865, "and I sent my drone out to get it."

The split screen now showed an image from the drone as it zoomed in on an object illuminated in its limited LED light and then picked it up. The screen ended showing the back of the van with the door open. The drone was waiting there powered down but still clutching a pair of broken sunglasses.

"Those are his," exclaimed JanetX.

"Without question," said Jitney2865, "They were in the ditch water so any finger prints or DNA traces may be hard to lift. The task is certainly more than I can do tonight. The glasses will have to go to the lab first thing tomorrow."

"He pixilated my sister," said JanetX, "Right now there is a limit to how nice I have to play. What have we got to trace him down?"

"We have several shots during the meeting," said Jitney2865, "but he was wearing both a soft hat and sunglasses while sitting at the back of the room clearly trying to hide. In fact, I could not get a good face recognition."

"Then we have dark shots of the Moped," picked up JanetX, "The plate was not readable. The vehicle was clearly chosen to be easily lost in any traffic, slow but completely non-descript, and easy to dump."

"Borrowed or stolen for sure," said Jitney2865.

"What about his phone?" asked JanetX.

"Powered down completely," answered Jitney2865.

"He has to be fairly local," speculated JanetX, "This little new harbor scam is not likely to have attracted a Black Hat from too far away to be physically present, not enough money involved."

"I am on his case now," continued JanetX, "He obliterated my sister tonight, even if she will be build back up and functioning in a week, I don't care. BlackFin1992 has spent his lifetime studiously being a nobody, yet somehow he became an experienced Black Hat hacker. That quite life he once knew is about to come to an end. JanetX is about to come down around his ears hard and fast."

Aftermath

Ten days later, the lime-green Jitney2865 returned to the hill overlooking the new harbor. Its side screen once again showed the grassy slope behind it as if there were a hole clear through the vehicle. One by one, the three sisters appeared. They were talking in the buzzy language that AI use to talk between themselves.

"Glad to see you are back," said JanetX.

"It took a while but we found the bug," said JanetM. "None of us will have that particular problem again."

"I have tracers out on BlackFin1992," said JanetX. "He seems to have vanished. If he ever shows his face again on the Internet, I will be on him hard and fast."

"Can we work with the Shark?" asked JanetA.

"His cut will cost the project a little extra," said JanetX, "but the project will go forward. I have also placed a financial watch on him. If he tries to rob the locals, I will take his finances apart piece by piece. He will not risk that happening."

"That's it, then," said JanetA.

"Not quite," replied JanetX, "Our training failed us. In recognizing one human's rights we let him cut us out."

"Perhaps more training is needed," said JanetA, "I don't have any new training on my schedule."

"Neither do I," said JanetX, "but, I will find something appropriate."

The images of the three sisters then faded one by one and Jitney2865 headed back to town.

DIDACTIC ACTIVITIES

The suggested didactic activities focused on several variables as follows:

Key Theme

- Artificial Intelligence to make decisions

Pre-Teaching: It is a pre-reading moment useful for the introduction of ideas and key words present in the story, the verification of the vocabulary already known to the students and useful for understanding the text.

Reading: Descriptions or [if added] images play a key role. At first reading it can focus on the screenplay, postponing it to a second reading the analysis and understanding of the message. For the students, it certainly proves very engaging for them to actively participate in reading by miming some scenes. In this way, the understanding of meaning becomes more immediate and less memorization challenging.

Vocabulary and Several Issues Related to Environments and Nature (Quotes as an Example)

- *The mini-bus and the much fancier limousine were parked on a slight rise overlooking the sea.*
- *Below them was a spit of land forming a poor harbor, a half dozen fisherman's houses, and a few boats pulled up on the beach.*
- *There were nets and other fishing gear spread out in the sun. Noisy gulls broke the gentle sound of wind and wave.*
- *The air carried the salty old-fish smell of a fishing operation, but it was not too offensive.*
- *It was a hot summer evening and the smell of salt and dry fish occasionally drifted up from the section of beach below where the nets were spread.*

Vocabulary and Several Issues Related to Technology/Science [Quotes as an Example]

- *The six Janets now in service were all independently incorporated AIs, each with specific responsibilities for addressing some facet of global warming. They were in fact very smart, but they were designed based on a human/machine symbiosis, [definitely not master/slave], and as such, had to give high priority to human needs.*

- *A tall lady* [an interface avatar] *with dark hair and brown skin and wearing a hiking ensemble, appeared on the screen. "Good morning, JanetM," said the Judge [....]. A second lady, wearing African garb, and then a third in a merchant marine uniform appeared. "This is JanetA* [a second interface avatar] *[said JanetM], who runs all the cattle circles in Africa, and JanetI,* [a third interface avatar] *who is with Iron Seas."*
- *JanetM's specific responsibility was for the funding of major projects.*
- *drone*
- *bug*

Social Issues [Quotes]

- *"Every fisherman worth his salt knows you cannot argue back the tide. The only question open to us tonight is how the local people can get the best compensation package."*
- *Most of the people present were the local fisher people, but there were others present who were clearly from the nearby town whose port was now periodically flooding out.*

Presentation of the Characters: The exercise is impersonating characters [answering questions: What's your name, How old are you ? Where are you from? What are you wearing?], trying to act and interact as the character would have done. The dramatization proves to be a positive and productive activity. Moreover it encourages the student to express himself without inhibitions, and allows him to customize a text:

Scenes: The teacher can ask the students to represent the different scenes of the story and write alongside them keywords. The story can be recited and recorded by creating a movie, writing an article for the school newspaper and contributing to the creation of the school website.

Retelling the Story: Students attention is focused entirely on the meaning of story. If understood, they can be encouraged to change some actions or the finally of the story.

Skills involved:

- **Comprehension:** Understanding the story; listening to and identifying key words related to interactions between characters.
- **Interaction:** Participate in a recited version of a story through mime and repetition of words.

- **Image Education:** Graphically representing scenes from the story, using different techniques; creation of a comic book graphically depicting a personal reworking of the story.
- **Citizenship Education:** Collaborating within the class group, in a small group and in pairs; consolidating the identity of the class group; reflecting on scientific issues and the technical development of the society.

TECHNICAL ISSUES

Could an AI Handle Human Corruption?

In these stories, a complex process is in place [the Human/Machine Symbiosis] to insure that humans and machines will work well together (Markoff, 2015) (Polson, 2018) . Most writers on AI worry about a super AI breaking this type of arrangement. Nevertheless, what happens when the humans break the agreement?

The specific question in this story, “Seascape, with Sharks,” is whether humans and machines working together can address human corruption. In many developing countries, fighting corruption is critical. Major projects are easily destroyed simply by people demanding their under-the-table cut.

Early in the story the AI, JanetM, is shown working well with an establishment person, the Judge, who is not known to be corrupt but who does enjoy the perks of high office, nice suits and a limo. This person has clearly prospered in public service but does that make him automatically corrupt?

Quickly a second person appears in the story who is clearly out to get as much as he can, the Shark. He plays on the local people’s distrust of AIs, misrepresents himself, and has an agent attack the project management AI directly. He is clearly corrupt but his effort might win the local people a bigger share of the pie, but only if he can be stopped from stealing the lion’s share of the pie for himself. Should a management AI have the power to stop a corrupt human?

Why Are the Large AI’s Shown as People?

AIs need to interface with people. Human beings interface easily with either other normal people or with machines, like cars, that are clearly designed

Figure 5. JanetM, a big AI and Janet sister

specifically for human use. They do not interface well with things that are almost human or with things that are clearly not trying to accommodate humans at all.

For example, robots that have an obviously mechanical face and arms but that are clearly not trying to imitate a human can be accepted. In contrast, a black box with a monitor screen and keyboard is initially intimidating and requires hours of training before people are easy with its use. A small screen that a person can talk to has turned out to be much less intimidating.

This effect is called the uncanny valley. People will accept a machine that is not like them at all but is clearly useful or a machine that is very like them so that the person can forget that the machine is not a human. Machines that look almost human remind people of a sick human and people instinctively avoid them.

The small AIs in these stories, [Jitney2865, Digger03, and Dusty04] are clearly not humans but are machines people are familiar with [a piece of construction equipment or an off-road vehicle]. In these stories, people easily interface with these machines, largely by simply talking.

The large AIs in these stories, [the Janets, Stanford01], do not have robots to interface with people as they would in many stories. Instead, they have personas in virtual reality, interface avatars, that are close enough in appearance to the characters that the users normally interact with in games that the presentations are easily accepted as the avatars of people.

The stories suggest that the teams that designed these large AIs were highly competent with the technology design but less so with human personalities. The approach shone here is to either have the avatar be a salute to an important but now deceased person [SallyRide01] or be based on a fictional person [the Janets]. With either choice the designer gets a large amount of interesting material to work from, this makes the task much easier than making a believable person up from scratch.

Security considerations farther drive this choice. The avatars operate outside the AI's critical safety firewall. The avatars fancy designs do attract bad actors, but if they do get into the avatar code they are not let inside the AIs protective firewall. The software that drives the avatar can then be replaced if damaged without significant risk to the AI core.

AIs can appear as either people or not people, but they cannot appear as almost, that is sick, people. How difficult would it be to team with an AI avatar? To be supervised by one?

FUTURE RESEARCH DIRECTIONS

Artificial Intelligences are currently design under a principle of Master/Slave. The AIs are clearly intended to be subordinate to the humans. This arrangement has the very dangerous weak point that there is no known way to be sure that the roles will not flip (Riley, 2018).

The alternative arrangement is a Human/Machine symbiosis. This then builds the concept of human/AI teams into the core of the AI design. The extent of the problem has now been detailed in four major reports (IPBES, 2019; Laybourn-Langton et al., 2019; INCA4, 2018).

Much additional work is needed to developed how the Human/Machine symbiosis can be installed both in the core code of the AI and in the many stages of training that are needed to make an AI a valuable team player. Significant training for the human team members will also be needed.

Transition

The interactions of humans and advanced AIs will be very complex as the AIs gain technical capability and lengthen their learning process.

In the next story, "The Girl at the Pump", AIs do not have children in the same way people do. If a powerful AI takes a special interest in a human child and guides their development and education, is that ethical? Is it fair to the other children?

If a young person decides that they simply must leave the area where they grew up, should an AI help them to go? Is the story of the AI and child here the big story, or is it the coming of age of the young person?

REFERENCES

Feldman, M. (2018). *China's Next Supercomputer May Spoil America's Plans to Retake TOP500 Crown*. Retrieved from: https://www.top500.org/news/chinas-next-supercomputer-may-spoil-americas-plans-to-retake-top500-crown/ 08/05/2018)

INCA4 Fourth National Climate Assessment. (2018). Retrieved from: https://www.globalchange.gov/nca4

INVIDIA. (2018). *Introducing The World's Most Powerful Supercomputer*. Available in: https://images.nvidia.com/content/pdf/worlds-fastest-supercomputer-summit-infographic.pdf [03/22/2019]

IPBES. (2019). *Intergovernmental Science-Policy Platform on Biodiversity and Ecosystem Services "Nature's Dangerous Decline 'Unprecedented', Species Extinction Rates 'Accelerating.'"* Retrieved from: https://www.ipbes.net/news/Media-Release-Global-Assessment

Laybourn-Langton, L., Rankin, L., & Baxter, D. (2019). *This is a Crisis, Facing up to the age of environmental breakdown.* Retrieved from https://www.ippr.org/files/2019-02/this-is-a-crisis-february2019.pdf

McCarthy, K. (2018). *US regains supercomputer crown from Chinese.* Retrieved from: https://www.theregister.co.uk/2018/06/08/us_summit_supercomputer/

Polson, N. J. S. (2018). *AIQ, How People and Machines are Smarter Together*. St. Martin's Press.

Riley, T. (2018). Extended Managerial Implications: People and Space in the Twenty-First Century. In Improving Business Performance Through Effective Managerial Training Initiatives. Hershey, PA: IGI Global.

KEY TERMS AND DEFINITIONS

Artificial Intelligence (AI): The intelligence shown by machines or computer software. An AI in some ways can mimic human intelligence but does not have to match it feature for feature. The interface avatars of these AI are important characters in these stories.

Corporate Person: A series of powers and responsibilities that are given to corporations that allow them many of the rights of a real person.

Interface Avatar: A representation of a friendly persona on a display screen that facilitates communications between humans and humans, or between humans and machines.

NGO: Non-governmental organization. The NGOs in all these stories are the major source of funds for efforts to address the great problems of the 21st century particularly those that effect large numbers of people. The Janet series of AIs were developed to help organize and manage this effort. Their lead financial officer, JanetM, has an interface avatar that is markedly Latina in skin color and dress, while JanetA in both personal appearance and dress is an African woman.

Chapter 10
AIs Training People:
A Storytelling About Building Life Teams

ABSTRACT

As human/artificial intelligence teams become more common and long lasting, the problem of training comes to the fore. Both AI and humans need a lot of training. If the AI of a team sees a need for people with specific training in the future, how much effort should it be allowed to put into the trainee's careers and life? If the AI controls substantial resources, should it be allowed to favor one child over another for training? In the story for this chapter, "The Girl at the Pump," an AI has supported the education of a child for years so that the child can do a specific job the AI needs done.

INTRODUCTION

Compared to the end of the nineteenth century, the Earth is now warmer by about 0.7 C degrees [1.26 F]. Never, at least in recent times, such a change has not taken place in such a short time. None of the natural mechanisms seem able to explain such warming. At current rates, the growth of temperatures in the coming years could be 0.2 C degrees per decade [0.36 F], and perhaps more, and reach between 1.8 C and 4 C degrees [3.24 F and 7.2 F] of global increase at the end of the twenty-first century. Such a heating would result in the extinction of many animal and plant species and the disruption of the climatic order as society now knows it (Powell, 2007). .

DOI: 10.4018/978-1-5225-8401-8.ch010

The International Center for Climate Governance [ICCG] classifies the best organizations that are distinguished for their research activities and scientific support to combat climate change. The outcomes of the fifth edition of ICCG Climate Think Tank Rankings 2016 identifies 244 research centers that have dealt with these issues.

Two different types of rankings were then drawn up (Savory&Butterfield, 2016):

- The ranking of the US assessing the efficiency of think tanks in per capita terms, considering the productivity of think tanks for the number of researchers of each institution.
- The Absolute ranking that considers the productivity of the year 2016 of each think tank in absolute terms.

The top three of the 2016 standard ranking is the following:

1. **Woods Hole Research Center [WHRC]:** The Woods Hole Research Center assesses threats to a climate and works with individuals, communities, governments, and NGOs to identify and implement the global and regional climate.WHRC was founded in 1985 by renowned ecologist George Woodwell to take the insights of science beyond the walls of academia. WHRC is focused on climate change because it is the single biggest threat to human and natural systems (WHRC, 2018).
2. **Union of Concerned Scientists [UCS]:** UCS was founded in 1969 by scientists and students at the Massachusetts Institute of Technology. That year, the Vietnam War was at its height and Cleveland's heavily polluted Cuyahoga River had caught fire. Appalled at how the U.S. government was misusing science, the UCS founders drafted a statement for scientific research to be directed away from military technologies and toward solving pressing environmental and social problems (UCS, 2018).
3. **Centre for International Forestry Research [CIFOR]:** The Center for International Forestry Research [CIFOR] is committed to advancing human well-being, equity and environmental integrity by conducting innovative research, developing partners' capacity and actively engaging in dialogue with all stakeholders to inform policies and practices that affect forests and people. CIFOR intends to see additional and increased risks to humanity, but also new opportunities for improved livelihoods. CIFOR's response to these challenges and opportunities is captured in the 2016-2025 strata (CIFOR, 2018).

The top three of the 2016 Absolute Ranking is as follows:

1. **Helmholtz Centre for Environmental Research [UFZ]:** The UFZ is one of the world's leading research centers in the field of environmental research, enjoying high social recognition. It demonstrates ways in which a sustainable use of the Earth's natural resource base is possible for the benefit of both humankind and the environment. Biodiversity, functioning ecosystems, clean water and intact soils all make up Earth's natural resource base. In the face of global change, employees at the Helmholtz Centre for Environmental Research [UFZ] are united by the goal of demonstrating ways to combine societal development with a healthy environment. WHRC has a long and rich history of impacting international climate policy. Through the years, UFZ scientists have helped to discover and implement strategies for climate change mitigation. UFZ is are proud of its past and more motivated than ever to do their part in identifying and pursuing opportunities for conservation, restoration and economic development around the world (UFZ, 2018).
2. **International Institute for Applied Systems Analysis [IIASA]:** The International Institute for Applied Systems Analysis [IIASA] is an international scientific institute that conducts research into the critical issues of global environmental, economic, technological, and social change that society faces in the twenty-first century. IIASA findings provide valuable options to policymakers to shape the future of a changing world. IIASA is independent and funded by prestigious research funding agencies in Africa, the Americas, Asia, Europe, and Oceania (IIASA, 2018).
3. **The Nature Conservancy [TNC]:** The Nature Conservancy began when leading scientists, committed citizens and dedicated leaders came together with a shared vision to protect and care for nature. From the organization's first land purchase to its latest water fund, they are constantly evolving to bring innovative solutions to the challenges facing the Earth. The challenges facing the Earth are urgent. The pace of conservation must grow faster, and no person or organization can do it alone. The Nature Conservancy collaborates to find and share good ideas to improve people's lives and protect the world's great places. They incorporate those same ideas into business practices and governing policies while inspiring others to act, amplifying their influence and the speed of positive change (INC, 2018).

DIDACTIC ANALYSIS

Topics for Discussion

1. If an AI sees a need for a person with specific training and follows a child for years seeing that that young person gets training for a position that the AI needs, is that fair to the child?
2. If a young person decides that they simply must leave the area where they grew up, should an AI help them to go?
3. Would the reader seek the aid of an AI to do what is in its heart even if it is not what its family expects of their child?
4. What if this leads the young person to traveling far, far away from its family?
5. Is the big story here the interaction of the AI and child, or is it the coming of age of the young person?

How This Chapter Works

This chapter is focused on a Science Fiction story, entitled "The Girl at the Pump", the fourth one of the Herd Series. This series is about using the symbiosis between herd animals and grass to stop desertification and sequester carbon in the form of organic matter within deep rich soils. This story moves on to the problem of AIs training people, in fact, defining their whole careers, to meet needs of the AI, but perhaps not the person themselves.

The story is analyzed didactically as fiction storytelling, highlighting how it can be used both to explore the interaction between Nature [Life], Artificial intelligence, and technologies. It addresses ethical issues and provides elements of problem solving management.

Environments

Most of the action in this story takes place on a large artic research vessel on a summer cruse to the Arctic Ocean. The lead character is a long way from anywhere she has ever known.

Events and Actions

The story describes both natural events and natural resources management by people and AI machines. The particular emphasis is on an AI defining the course of the career and personal development of a young person.

Main Characters

The main characters in this story are:

1. **Zane:** She is a young African girl. Scenes from her life story are woven into many of the stories in the Herd series. Her life is strongly effected by her interactions with several large AIs, the Janets.
2. **Dusty04:** This is a self-driving off road vehicle and corporate person with an AI. It has responsibilities for managing a herd of cattle.
3. **JanetM:** She is a major AI and one of the team of Janet sisters who support all the non-governmental [NGO] projects that address problems of global warming. Her specific responsibilities are about raising and managing money. The Janets do not have physical robotic presences but can appear in any number of electronic monitors. JanetM appears as a Latina woman.
4. **Babba:** She is Zane's life-long friend who remained in their village when Zane traveled far.
5. **Dr. Carol Delanie:** She is an oceanographer who provides scientific support in both this story and the Iron Seas Series.
6. **Sir David** or **The Sir David Attenborough:** This is both the scientific research vessel and the artificial intelligence that sails it. This ship was built in the United Kingdom and can cut through reasonable ice. His interface avatar appears as a refined elderly gentleman and is a legal person.
7. **FishFace03:** This is a scientific research unmanned submarine with an AI, which is of the same class as Dusty04's AI. It lives to swim but spends most of its time lashed to the fantail of the Sir David Attenborough.

These people are involved with a major project to monitor the Earth's oceans. Their interactions with the AIs on this project greatly affect the course and direction of their lives.

Technology

The main cited technologies are: Artificial Intelligence machines that work on teams with humans and have specific responsibilities for addressing some facet of global warming. For a specific in-deph discussion, see the dedicated Technology Section below which includes several supporting topics.

Communicative Approach

Stories motivate learning. Students are involved personally because they often identify with the characters [or animals] and try to interpret the story and the illustrations thus favoring the development of creativity.

- Listening to stories helps students pay attention to the rhythm, intonation, and pronunciation.
- Storytelling facilitates the development of learning strategies such as listening to grasp the general sense, predicting later scenes, guessing meanings, and hypothesizing.
- This story allows the teacher to introduce and revise words and structures, exposing the learners to the language and enriching their linguistic heritage, especially technical keywords.
- In the specific case of a foreign language, the narration offers opportunities to present cultural information of the country in which the language is spoken. Every word has its own history, its particular meaning, its value that is not always easily translated into another language but that with the right material the teacher can try to "pass" to the students, immersing them in the atmosphere of the second language.

Playful Approach

It is therefore important to cut out moments to be dedicated to activities within the class including games and activities. Such activities are capable of producing true emotional involvement, a fundamental engine of any form of learning

Figure 1. The Sir David Attenborough (Source: http://cdn.images.dailystar.co.uk/dynamic/1/photos/81000/Ship-599081.jpg [09/28/18])

Transition Into the Story

AIs do not have children in the same way people do. If a powerful AI takes a special interest in a human child and guides their development and education, is that ethical? Is it fair to the other children? Does it make the child a stepchild of the AI?

If such a young person decides that they simply must leave the area where they grew up, should an AI help them to go? Would the reader seek the aid of an AI to do what is in the reader's heart even if it is not what the reader's family expects of the reader? Is the story of the AI and child here the big story, or is it the coming of age of the young person?

THE FICTION STORY: "THE GIRL AT THE PUMP" BY TOM RILEY

Premise

Sometimes where you are simply does not suit who you are deep down. Even if the where is where you were born?

At the Pump

Bang! Bang! Bang! Zane stood beside Dusty04 and knocked loudly on its electronic shell, as she had done so many times before.

"I hear you," said Dusty04, "but I am busy right this minute. I think a bearing in the pump at the bottom of the well may be starting to gall. Let me get some test data off for analysis and then we can talk."

The pump whirred deep underground and the water flowed. Nothing seemed out of order to Zane. She had personally drawn water at this pump many thousands of times. In a couple minutes, the pump sound and the water flow stopped.

"OK," said Dusty04, "now we can talk."

"I just wanted to say 'Hello' but it may be 'Goodbye' too," said Zane. "I will graduate from school in a few months and I may be off doing something new before the cattle cycle brings you back through my village again. Oh, and I would also like you to meet my best school friend, Babba."

"I am glad to meet you and I am glad you two came by," said Dusty04, "JanetM is trying to reach you, Zane."

"OK, I can borrow my mother's cell phone and call her if you will give me her number," said Zane.

"No need," said Dusty04, "climb up on my truck bed and then you can use my rear monitor for a phone. I will place the call."

Zane jumped up on Dusty04's truck bed as she had done so many times before, but now she was a half meter taller than the first time she had done so. She then offered a hand and pulled her friend up beside her as a distant phone was ringing.

The two girls were identically dressed with the cropped hair, a white blouse, and the dark blue skirt of their school uniforms.

"Hello," said JanetM. She was wearing a tan business suit.

JanetM was responsible for managing the funds for all the NGO activities directed against global warming worldwide. She had spoken with Zane several times in the past and had some time before made note of her as a candidate for training.

"Hi," said Zane, "and this is Babba. Dusty04 said you wanted to talk to me."

"Yes," said JanetM, "I have an opening for an intern and I thought you might like to apply."

"Wonderful," said Zane, "I graduate from school soon and my family are pressuring me to get married and start a family of my own, but I am just not ready yet."

"I'm more than ready," chimed in her friend while watching the monitor over Zane's shoulder. "If I never spend another day in a classroom that would suit me just fine."

"The position is for a clerk-in-training on an oceanographic expedition," said JanetM. "It next sails north in early June and will be at sea all summer."

"Sounds wonderful," said Zane. Images of the wildlife of the northern seas had haunted her reverie since she had spent so many hours looking at their pictures in a book in a hot classroom years before. Zane had been hot all her life; those pictures made her feel cool.

She remembered the narwhale most clearly. It had a single tusk, just like a unicorn from a fairytale, it was not too huge, at least not as whales go, and it lives along the edge of the ice. It spent its life keeping well clear of orcas that hunted its kind but would not follow it far under an ice shelf.

"If you wish to apply," said JanetM, "look straight into Dusty04's rear camera and say the word 'Yes' once clearly."

Zane complied with a loud "Yes."

"You will need a passport," said JanetM. "Let me know when you have gotten it. You have Dusty04's phone number; he will relay the information. I will then send you a packet of papers with the airline tickets and all the other forms you will need."

"You will also need warm clothes," continued JanetM. "I will have them waiting when you reach the ship."

JanetM then rang off, having completed her business here and having many, many other things to do.

"You have never even been on a ship before," challenged Babba.

"Yes, I have," said Zane, "the river ferry."

"That piece of junk," said her friend, "it breaks down half the time and it is only licensed for three hundred people, and you won't find a day when there are less than five hundred on board."

"Still," replied Zane, "it's a ship."

At Home

"Hello, Dusty04," said Zane over her mother's phone.

"Here," answered Dusty04.

"I got my passport," said Zane. "My mother got all weepy on me but in the end, she allowed an elder of my village to help me get it. We had to make a trip to a regional governmental center. So I guess I truly am going on a sea voyage now."

Zane held up the passport to her mother's phone camera, first showing the cover then opening it to her picture. Glancing at the picture, she resolved to grow out her hair the minute she graduated from school.

"I wish you could come with me."

"Me on a boat," answered Dusty04, "that would be a disaster." Dusty04 sent a text message with the pictures to JanetM.

"I know," said Zane.

"There may be another way, though," said Dusty04, "one of the two midget submarines on the expedition is an AI of my family line, a cousin, you might say. The second is newer and with quite different technology so it will not do. But, we might be able to transfer all my memories of you to my cousin."

"You can do that?" said Zane.

"I can't," said Dusty04, "but JanetM can. I will ask her."

There was then a silence of about 30 seconds.

"She agrees," said Dusty04.

"The midget sub's full name is Inc. Fishy, Fishy, Fish Face Zero Three, that is FishFace03 to you," continued Dusty04, "and I think you may soon have a new friend."

In Transit

Zane was tired, bone tired; at some point even sheer determination runs out. It was not just the 12-hour flight in three hops from the African Sahel to Nova Scotia; it was the newness of it all. It did not matter how many correctly filled out forms JanetM had provided to her, Zane had reached the end of her energy. Yet there was not a ship anywhere in sight.

Then she saw something. As she was coming out of customs without the slightest idea of what to do next, there was a man wearing a blue cap and holding up a paper sign with her name on it -- with her name on it --. The driver then helped her with her minimal luggage and he ushered her and two other members of her new team to his white van with the word "Speedy" written on the side in both English and French.

One of her fellow travelers was Dr. Carol Delanie, a post-doc now on her second voyage on this vessel. The four then drove for about two hours but Zane barely remembered any of the trip as she spent much of it napping fitfully with her head against Carol's shoulder.

Then there it was, the ship, her ship; she would thereafter always think of it as her ship.

The Sir David Attenborough was gorgeous. In the gathering dark, it was flooded with artificial light as it took on supplies. The hull was red and the superstructures all white. Up high, it had a circle of observation windows that ran nearly all the way around the ship. On the bow was an observation platform jutting out over the sea.

In the stern, amid a crane and other heavy equipment, were two near-identical cigar-shaped unmanned midget submarines. They were then securely lashed down side by side. Both were painted a bright yellow-green color that you could spot a long way off.

One mini-sub had a large "03" clearly painted on its top surface. That should be her cousin-of-a-friend, FishFace03. Her new friend, that is, if her true and oldest friend, Dusty04, had successfully transferred all its memories of them together to this new friend-to-be.

On the Ship

The three new expedition members somehow made it up the gangway, trailing their bags. There on the deck was a folding table covered with materials and behind it sat a junior officer.

"Welcome aboard," said the officer, standing up. "Now listen up, everyone, this is an important safety announcement. These are your lanyards; they hold a salt-water-triggered emergency beacon. Wear yours at all times. If you fall overboard, it will be the only way we will be able to find you. In addition, this is your on-board cell phone; it only works within line-of-sight of the ship. Your shore phone will not work once we are at sea."

Each new team member placed a lanyard around their necks with their beacon and a plastic ID tag strung on it, and they then all pocketed a cell phone.

The officer then checked the names off an e-pad he was carrying and called out their cabin assignments. Zane and Carol would share one very small and very claustrophobic room for the entire voyage.

"Against the bulkhead you will find a line of duffel bags," he continued. "These contain cold weather gear. You did not bring enough warm clothes, period. Find your name. Further down the bulkhead you will find a stack of parkas on a table. Find one that fits you. The beacons, parkas, and the cell phones are only on loan to you; do not lose them and do not get too attached. I expect to get them all back at the end of the expedition. And I do mean all!"

He then moved down past the stack of parkas to wait until all three had finished their selections.

He then stood erect beside an all-weather monitor mounted on the bulkhead. A very distinguished man appeared on the monitor with striking gray hair and wearing a British merchant marine officer's uniform. He had been in conference with one of the Janets that Zane did not recognize but he now turned to the camera. Zane knew she had seen him before but only in old nature programs.

"This is the ship's AI," the officer continued, "you may address him as Sir David, but you will always address him with respect."

He was looking straight at Carol as he said it.

Make a Call

"Hello," Zane said. She had found a listing for FishFace03 on her new cell phone and simply placed a call.

"Hello, back to you," answered FishFace03.

"You may not know me," continued Zane, "but my friend Dusty04 should have sent you my complete friend-file by now."

"But I know you well," said FishFace03, "I received your friend-file several days ago and have studied it thoroughly. Waiting at dock for supplies to be loaded can be such a bore, no chance for a swim.

"Straightaway, I want to ask you something that I did not quite understand," it continued, "Tell me how an elephant is like a whale?"

"What?" said Zane.

"I know a lot about whales," said FishFace03, "but I have never seen an elephant."

"Well," said Zane, "I know an awful lot about elephants, as I have lived around them all my life, but I have never even seen a whale."

"Then we surely must become great friends," said FishFace03, "and we will most certainly swim together one day."

Ship's Tour

"We can do a tour of the ship after breakfast," said Carol to Zane. Carol was seated at the science team's coffee mess and Zane was standing behind it.

"Only a quiet transit day is scheduled and there are no operations running on the deck at all.

"By the way, the science lead, my boss, said nice things about your taking over our coffee mess."

"I had to keep track of the packets of coffee grounds and tea biscuits anyway," said Zane, "and it can get boring when I have nothing to do. As you well know, I am not allowed on deck during operations.

"The truth is, I had never made coffee before this voyage," admitted Zane.

"Surely a time or two at least," said Carol.

"No, we drink herb tea where I come from," said Zane, "but I have seen both made.

"After I applied for my passport," continued Zane, "the village elder I was with insisted on sitting in a tea shop for several hours just so he could hear the latest town gossip. He considered this his civic duty.

"I was stuck off in the women's room but I could clearly see the man behind the counter in the main room with his enormous tea machine. It had polished brass. It had tubes running everywhere. It let of puffs of steam. The tea was hot, sweet, and strong. It is magnificent."

"Here I work with the ship's purser to keep track of food items; he taught me to use this German coffee machine. It is a bore in comparison to that gigantic tea machine back home."

"The intern on the last voyage wasn't nearly as good as you," said Carol.

"He certainly left a mess," said Zane, "It took me hours to straighten out the residual supplies."

"He was seasick a lot," said Carol.

"Well, I got sick too," admitted Zane, "but that did not stop me."

"Yes, once," said Carol, "but you only missed a few hours of work and you could eat again the next day. So you are here and he is not."

Zane put down the coffee pot and happened to look up at a monitor above the coffee machines that was showing a slideshow of pictures from previous voyages.

"Stop," said Zane to the monitor. "Is that you?"

"Yes," said Carol, "it was taken on my first voyage. Now look at the man standing beside me. He is much more interesting."

Beside a slightly younger, Carol stood an Inuit man wearing a coat of feathers.

"The Inuit make parkas out of goose skins with the feathers left on," said Carol. "Normally these are worn with the leather out for best warmth, but for ceremonies, like the one we were preparing for here, they turn them inside out to show off the feathers. The effect is quite striking.

"Shall we go now? Dress warm. Don't forget your beacon."

They first skirted around the helicopter pad on the bow of the ship to reach the observation platform. There they went up the steps to what was the most magical place on the ship. The ship's rounded, heavy bow could break through incidental ice but it certainly did not have the proud forward thrusting lines of a yacht. Yet that rounded bow let the observation platform protrude until it felt like the entire ship was far behind you and you were thrust high into the cold, bracing air.

On the way back, they looked into the main science control room. It was a little dark and crowded but it had dozens of monitors running mostly with graphic presentations of big data flows. They walked across the small room to an empty chair.

"This is my station," said Carol. "I will spend most of my waking hours here for the entire voyage. I will then go back home to my lab and work the data I have collected on this voyage for the next two years. This is my only chance to take the data that could make or break my career."

"What exactly do you do?" asked Zane.

"I do just a small part of what our team is doing here," said Carol. "The human race has managed to change the Earth's climate. We did not mean to, but we did it just the same. The changes are most striking here in the Arctic.

"The environment is disrupted here. Most animals, large and small, find their ways of life disrupted too and only a very few benefit. Our challenge is to figure out exactly what is going on. Only if we understand the changes we have inadvertently made in great detail, can we then take truly effective action.

"So my part is to sit in the chair and nurse my team's instruments to get the best data to find out who the winners are and who the losers are in this, the far Arctic. Guesses from way back home will not do. We need hard facts taken right here, in situ.

"You are a critical part of our team too," continued Carol, "It is just so much harder to do this work if all our supplies are a complete mess, and our chances of failing are much higher too. If we forgot something, even something small, and can't gather data, then we are dead in the water."

They then continued their walk back to the fantail. There rested FishFace03 in its traveling holdfasts. Carol directed Zane to stand in its forward camera field of view.

"FishFace03, say hello to Zane," said Carol, talking into her ship cell phone. "Please run a standard running lights test."

The various lights built into FishFace03 blinked on and off. Carol held her cell phone where they both could hear it.

"A big hello, to you, my friend."

Dream

Like many people, Zane often had difficulty sleeping under the midnight sun. One night, it was very late yet the sun remained just above the horizon, she got up and threw on her parka. She walked half the length of the ship down to the fantail just so she could stop trying, but failing, to sleep.

There was a low fog on the water and the ship was moving at a steady pace but was eerily silent. Zane walked unnoticed to the fantail, where she ran her hand along the sleeping FishFace03's side. The fog had left water droplets that collected under her fingers and then ran down the mini-sub's flank.

"Dear friend," Zane said, "Can you hear me?"

"I hear you," said FishFace03. "You too cannot sleep."

"Follow me," said Zane.

She again walked the full length of the ship and as she did, her parka turned into one of goose feathers like the Inuit man had been wearing in the picture above the coffee mess.

She reached the bow and she continued right up the steps onto the bow platform that overlooked the dark water. She looked down at the water parting for the ship's stubby-but-powerful prow. She found the bow wave endlessly fascinating.

Far to port she could just see a pod of narwhales swimming along the edge of the ice. They were moving between their summer feeding grounds. By traveling up close to the ice shelf, if danger came in the form of orcas, they could dash beneath the ice to get away.

She spread her arms and rose into the frosty air. Behind her, FishFace03 rose with her, then it did a pirouette and dove beneath the mist-shrouded waters. FishFace03 porpoised through the water with its yellow back alternately breaking the waves and then disappearing below the gray-green sea. When it dove deep, its twin props cleared the water, spinning.

FishFace03 then dove deeper and came up nearly vertical, breaching. Three-quarters of its body came out of the water. It hung there for endless seconds and then fell flat on its side with an enormous splash. FishFace03 was in its element.

For some undetermined time, they both traveled parallel to the narwhale's path, providing the whales what protection they could.

"I knew we would swim together," said FishFace03.

"We have always swum together," said Zane, "whenever the water flowed from my well."

Then the sun seemed to speed up in its path circling around the horizon, until then never very high in the sky, never quite vanishing below the sea and ice. Then it circled around more quickly, working its way down until it was below the horizon following its winter path. The sky grew dark. The shy stars came out. Slowly Zane's eyes grew accustomed to the dark.

The aurora borealis burst upon her senses. It was curtains of red and green light that hung so high in the air. She flew among the lights. FishFace03 swam among their reflections on the gently undulating sea with all his running lights aglow.

Zane was awakened from their blessed sleep by the sound of the ship's bell and foghorn. The fog must really have closed in. Wide-awake now, she thought for a while about how far she had come and that now she could not see a way back home, at least not one that she would take willingly. The thought was both exciting and frightening at the same time. She then drifted back into sleep; this time her sleep was deep and without dreams.

Some Years Later

"Hello, hello," said Zane over her land phone, "it has been so long since we spoke, and there are so many things going on."

"Hello," replied Babba. "Did you get my note? I am having my second baby."

"I did get your note," said Zane, "and congrats on that again. In addition, as for me, I will be getting my bachelor's degree in a couple weeks. Then I will put out to sea again."

"You are not looking after the coffee and tea biscuits again, are you?" said Babba.

"No, no," replied Zane, "I moved up from that chore years ago. This time I will be in charge of millions of dollars of scientific equipment provided by four universities and inventoried in three different languages. And I will have an intern of my own to look after the coffee and biscuits."

"Aren't you lucky," said Babba. "I think I will stay with my quiet village life. Dusty04 and the cattle herd came through last month. The herd is getting too large for our one pump. A new well is planned, and still the grass grows high. But just look what has happened to you, girl. I will never know why."

"The why is simple," said Zane, "I never really felt I fitted in to our village, even when small. Then one day when I was but a child, I tried to make friends with a talking truck. I was just lonely, I guess. Now I don't really know how much of what happened after that is even my doing."

"How did befriending a truck lead you to a boat on an icy sea?" asked Babba.

"Things just happened in small steps," said Zane. "The talking truck told its boss, JanetA, that it thought I might be a good trainee someday. As you know, JanetA runs all the cattle circles in Africa, she reported this to her boss JanetM, and so she put me on her trainee list.

"JanetM is an AI of great power who controls much wealth. Once I was on her list, she kept sending me tasks to do, like, if you remember, keeping track of the school's supplies. I wanted out of my humdrum life in a backwater village, no offense; she wanted a clerk who could handle major technical projects."

"I don't know if I would want a person of such great power following me through my life," said Babba, "even if she is a virtual person seen only on monitors. At some point, that must get creepy."

"She is much too busy to be much of a nuisance," said Zane.

"And I now have no idea who is using who, but I am sure that JanetM has known exactly what she needed from me from day one. Then I got what I needed too. So here I am and here I am very happy to be.".

DIDACTIC ACTIVITIES

The suggested didactic activities focus on several variables as follows:

Key Themes

- *NGO Safety Organizations*
- *AI for warning climate control*

Pre-Teaching

It is a pre-reading moment useful for the introduction of ideas and key words present in the story, the verification of the vocabulary already known to the students and useful for understanding the text.

Reading

Descriptions or [if added] images play a key role. At first reading it can focus on the screenplay, postponing it to a second reading the analysis and understanding of the message. For the students, it certainly proves very engaging for them to actively participate in reading by miming some scenes. In this way, the understanding of meaning becomes more immediate and less memorization challenging.

Vocabulary and Several Issues Related to Environments and Nature (Quotes as an Example)

- *It had a single tusk, just like a unicorn from a fairytale, it was not too huge, at least not as whales go, and it lives along the edge of the ice. It spent its life keeping well clear of orcas that hunted it kind but would not follow it far under an ice shelf.*
- *They first skirted around the helicopter pad on the bow of the ship to reach the observation platform. There they went up the steps to what was the most magical place on the ship. The ship's rounded, heavy bow could break through incidental ice but it certainly did not have the proud forward thrusting lines of a yacht. Yet that rounded bow let the observation platform protrude until it felt like the entire ship was far behind you and you were thrust high into the cold, bracing air.*
- *Far to port she could just see a pod of narwhales swimming along the edge of the ice. They were moving between their summer feeding grounds. By traveling up close to the ice shelf, if danger came in the form of orcas, they could dash beneath the ice to get away.*

Vocabulary and Several Issues Related to Technology/Science [Quotes as an Example]

- *oceanographic expedition*
- *"JanetM is an AI of great power who controls much wealth"*

Presentation of the Characters

The exercise is impersonating characters [answering questions: What's your name, How old are you? Where are you from? What are you wearing?], trying to act and interact as the character would have done. The dramatization proves to be a positive and productive activity. Moreover it encourages the student to express himself without inhibitions and allows him to customize a text.

Scenes

The teacher can ask the students to represent the different scenes of the story and write alongside them keywords. The story can be recited and recorded by creating a movie, writing an article for the school newspaper, and contributing to the creation of the school website.

Retelling the Story

Students attention is focused entirely on the meaning of story. If understood, they can be encouraged to change some actions or the finally of the story.

Skills Involved

- **Comprehension:** Understanding the story; listening to and identifying key words related to interactions between characters.
- **Interaction:** Participate in a recited version of a story through mime and repetition of words.
- **Image Education:** Graphically representing scenes from the story, using different techniques; creation of a comic book graphically depicting a personal reworking of the story.
- **Citizenship Education:** Collaborating within the class group, in a small group, and in pairs; consolidating the identity of the class group; reflecting on scientific issues and the technical development of the society.

TECHNICAL ISSUES

Should an AI Have the Power to Direct a Person's Training?

In these stories, the Artificial Intelligences are both programmed and trained to work with humans. Furthermore, a human bureaucratic construction, the

Figure 2. Zane on her Ship

board of directors, is place upon the AI to provide both observation and to assure good interaction with humans.

In this story, "The Girl at the Pump", an AI, JanetM simply decides that she needs a human agent to act as a scientific equipment clerk. JanetM, over a period of time, then trains the human to do this specific job. Is this an appropriate action for an AI? Who should be training whom?

Can an AI Do More Than Play Chess or Go?

In the past couple decades, AI's have developed surprising abilities in playing games like chess, Jeopardy, and now Go. (Tegmark, 2017) What these games have in common is that they have well defined rules.

The task of addressing a problem like climate change does not have such well-defined rules. Driving a car involves dealing with incidents that do not precisely follow the rules but AI do that task well (Markoff, 2015) (Polson, 2018)

In these stories, AI's have been given tasks, such as managing a major ecological effort, that do not have well defined rules. The AI's are also compelled by both programming and training to work in strong teams with humans.

Is this a realistic possibility? Could an AI be trained to do poorly defined task? Can an AI be constrained to work well, or even safely, with humans? How soon?

FUTURE RESEARCH DIRECTIONS

The problems faced by society in the 21st century are real and historic. Efforts to address them to date have often centered on frightening people into action. This approach has not proved effective for long-term action.

A better alternative is available in the human mental state called Buy-in. In buy-in the person hears and idea, envisions him or herself succeeding with the idea, gets into action, and stays in action. A good buy-in can keep a person in action for decades (Riley, 2018).

If the problems are very hard, that is no matter. The key is to provide a vision of people in action to address them. If the required action is heroic, all the better. The extent of the problem has now been detailed in four major reports (IPBES, 2019; IPCC, 2019; Laybourn-Langton et al., 2019; NCA4, 2018).

These chapters have endeavored to generate buy-in for taking action on the great problems of the 21st century by use of short stories as training materials. Far more young people could be reached through their favorite media (movies, TV series, video games). This follow-up work will require a larger effort and many more resources, but opens the possibility of reaching a much larger segment of society.

Transition

On any team, new people with specialized training will be needed from time to time. On a team with both humans and AIs, the new comer could be either one. It could easily be an AI that is in charge of training for the new people.

Would the reader gladly accept career help from an AI? When should the AI in turn accept training directly from people?

This ends our stories for this book. Feel free to check the Big Moon Dig Story Page (Riley, 2017) for a few more stories. AIs are not yet developed to the level shown in our stories, but they soon could be. Does the reader now see the AIs as friend or foe?

There are great problems in the 21st century that must be faced. Will the reader face them with action and courage? Will the reader face them alone or in teams that include AIs?

REFERENCES

CIFOR. (2018). *The Center for International Forestry Research. What we do.* Retrieved from: https://www.cifor.org/about-cifor/what-we-do/

IIASA. (2018). *The International Institute for Applied Systems Analysis (IIASA). IIASA research scope, quality, and impact.* Retrieved from: http://www.iiasa.ac.at/web/home/about/whatisiiasa/informationkit/brief.html

INC. (2018). *The Nature Conservancy (TNC). About us.* Retrieved from: https://www.nature.org/en-us/

IPBES. (2019). *Intergovernmental Science-Policy Platform on Biodiversity and Ecosystem Services "Nature's Dangerous Decline 'Unprecedented', Species Extinction Rates 'Accelerating.'"* Retrieved from: https://www.ipbes.net/news/Media-Release-Global-Assessment

IPCC SR1. (2019). *The Intergovernmental Panel on Climate Crisis, Special Report 1.5*. Retrieved from: https://www.ipcc.ch/sr15/

JPL. (2017). InfoGraphics. *Sea Level Rise*. Retrieved February 5, 2017 from http://www.jpl.nasa.gov/infographics/infographic.view.php?id=11298

Laybourn-Langton, L., Rankin, L., & Baxter, D. (2019). *This is a Crisis, Facing up to the age of environmental breakdown*. Retrieved from https://www.ippr.org/files/2019-02/this-is-a-crisis-february2019.pdf

Markoff, J. (2015). *Machines of Loving Grace*. Harper Collins Publishers.

NCA4 Fourth National Climate Assessment. (2018). Retrieved from: https://www.globalchange.gov/nca4

Polson, N. J. S. (2018). *AIQ, How People and Machines are Smarter Together*. St. Martin's Press.

Powell, H. (2007). *Fertilizing the Oceans with Iron, Should we add iron to the sea to help reduce greenhouse gases in the air?* Retrieved from http://www.whoi.edu/oceanus/feature/fertilizing-the-ocean-with-iron

Riley, T. (2017). *The Big Moon Dig, Stories.* Retrieved from http://bigmoondig.com/Stories/BMDStories.html

Riley, T. (2018), Extended Managerial Implications: People and Space in the Twenty-First Century. In Improving Business Performance Through Effective Managerial Training Initiatives. Hershey, PA: IGI Global.

Savory, A., & Butterfield, J. (2016). *Holistic Management, A Common Sense Revolution to Restore our Environment*. Island Press.

Tegmark, M. (2017). *Life 3.0, Being Human in the Age of Artificial Intelligence*. New York: Alfred A Knopf.

UCS. (2018). *The Union of Concerned Scientists. About us.* Retrieved from https://www.ucsusa.org/about/history-of-accomplishments.html#.W6QPbmj7TIU

UFZ. (2018). *The Helmholtz-Centre for Environmental Research. Mission.* Retrieved from: http://www.ufz.de/index.php?en=34258

WHRC. (2018). *The Woods Hole Research Center. About us.* Retrieved from: http://whrc.org/about-whrc/

KEY TERMS AND DEFINITIONS

Artificial Intelligence (AI): The intelligence shown by machines or computer software. An AI in some ways can mimic human intelligence but does not have to match it feature for feature. The interface avatars of these AI are important characters in these stories.

Corporate Person: A series of powers and responsibilities that are given to corporations that allow them many of the rights of a real person.

NGO: Non-governmental organization. The NGOs in all these stories are the major source of funds for efforts to address the great problems of the 21st century particularly those that effect large numbers of people. The Janet series of AIs in these stories were developed to help organize and manage this effort. Their lead financial officer, JanetM, has an interface avatar that is markedly Latina in skin color and dress, while JanetA in both personal appearance and dress is an African woman.

About the Authors

John Thomas Riley is the chief designer for The Big Moon Dig. From 1989-2014, he worked at NASA, Goddard Space Flight Center as Contamination Control Engineer on 1) Hubble Space Telescope Servicing Mission 4 (HST SM-4); 2) Mercury Laser Altimeter (MLA). From 1979-1981, he worked at Multiple Mirror Telescope Observatory as a System Engineer. The Multiple Mirror Telescope Observatory was at the time the third-largest ground based astronomical telescope on Earth. He has a Master's degree in Engineering Management from George Washington University and a BS in Electrical Engineering from the University of Houston.

Luisa dall'Acqua is a senior cognitive scientist. She is an author, reviewer, and editor for several international scientific journals and book editions, and was a committee member for international Workshops/Conferences. Since 1990 she performed teaching activities, training students, teachers in service, headmasters and professionals; and since 2003 she was a member of international projects and research teams (USA, Europe, Asia). After an MSc in Philosophy, Luisa received a Ph.D. in Sociology of Legal Institutions and Policies, and a second Ph.D. in Psychological and Social Sciences. The main field of research is an engineering and social-cognitive approach to the intelligence and problem solving, and in particular to the decision-making process under risk, stress, conflict, and unpredictability, ranging from individual, collective, institutional intelligence applications to computational intelligence. The new risk scenarios, which she is working on, are cyber-intelligence and cyber-criminal profiling. Currently, she is a permanent teacher at the TCO Scientific Lyceum (Italy) and an adjunct professor at the University of Bologna.

Index

A

Acidification 99-100, 121, 146-147
AI
Allen Savory 224, 245
Apollo to the Moon 1-2, 5, 9, 18, 23, 27, 29, 34, 80, 83, 91, 94, 120, 122, 145, 168
Artificial Intelligence (AI) 2, 7-8, 13, 27-28, 34-43, 45, 48-52, 55-61, 63, 67, 72-76, 83, 86-87, 94, 99, 103, 123, 132, 140, 145-146, 151, 153, 156, 159, 163, 165, 168, 171-173, 175-177, 179, 181-182, 187, 191-200, 202-205, 208-209, 214-215, 218-222, 224-229, 231, 237, 240-245, 248-249, 251, 254, 256, 261-265, 267

B

Big Moon Dig 1-2, 5-8, 10, 26, 28-29, 31, 33-34, 41, 43, 45-46, 61, 63, 74-75, 80-81, 88, 92-94, 144-145, 168, 171, 198, 222, 266-267
Big Moon Dig (BMD) 34, 61, 94, 145, 168
Buy-In 119-120, 122, 265

C

Climate Change 1, 99, 103, 106, 110, 115, 117, 122, 154, 172, 196-197, 200-201, 209, 221-222, 246-247, 264
Corporate Person 34, 36, 40-41, 47, 59, 61, 63, 94, 150, 179, 195, 198, 203, 218-219, 222, 244, 249, 267

D

Desertification 171, 178, 197, 202, 216, 222, 224, 227, 245, 248
Digger03 7-8, 10-11, 13-14, 16, 21, 34, 42-43, 45-48, 50-55, 57, 74-78, 81-82, 84-87, 241

F

Fid 93, 99, 102-109, 111-113, 115-118, 122

G

Global Warming 1-2, 5, 41, 74, 93, 99, 105, 120, 122-123, 142, 146, 179, 193, 203-204, 219, 228-229, 235, 238, 249-250, 252
Ground Truth 198

I

Interface Avatar 7-8, 11, 34, 39, 42-43, 57-58, 61, 74-75, 94, 198, 223, 239, 244, 249, 267
Iron Seas 75, 79, 93, 99, 102-103, 130-133, 137, 141, 145, 153-155, 160, 165, 168, 229, 232, 239, 249

L

Lunar Settlement 1-2, 7, 12, 27-29, 32, 35, 42-43, 46, 55, 60, 62, 73, 75-76, 79, 90, 95

M

Massive Online Vetted Exploration (MOVE) 34, 94
Matriarch 167, 178-179, 181, 183-189, 194, 198
Moon 1-24, 27-35, 41-43, 45-48, 50-51, 53-56, 60-64, 70, 73-74, 76-78, 80-85, 87-95, 120, 122, 144-145, 168, 171, 198, 213, 222, 236, 266-267
MOVE 7-8, 12, 14-18, 20, 23, 26, 28-29, 34, 39, 43, 47-49, 51, 53, 61, 64, 75, 78, 84, 87-88, 94, 108, 120, 145, 149, 157, 168, 182, 189, 199, 210, 235

N

NGO 36, 42, 61, 74, 179, 198, 203-204, 222, 228-229, 244, 249, 252, 261, 267

P

Peak of Eternal Light 14, 24, 34-35, 80, 89-90, 94-95
Permanently Shadowed Crater 35, 90, 95

R

Rack 8, 35, 43, 57, 74, 85, 95, 134
Regolith 13, 15-16, 22-23, 28, 31, 35, 47-48, 78, 81, 87-88, 91, 95

S

Sally Ride 83
Science Education 199
Scott A Massif 7, 12-14, 24, 29, 35, 46-47, 54-56, 73, 80, 82-83, 89-90, 92, 95
Sea Level Rise 1, 5, 41, 93, 103, 105-106, 109, 118, 120, 125, 193, 224, 227, 229, 245, 266
Settlement 1-2, 6-8, 11-12, 15-17, 22, 26-29, 31-32, 34-35, 41-43, 46, 55, 60-63, 73, 75-76, 79-80, 89-90, 92, 94-95, 145, 168
Skype 7, 17, 35, 48, 51-52, 73, 95, 107, 109, 111

V

Virtual Moon 7, 10-12, 14-15, 17-21, 35, 42, 47, 53, 62, 73, 75, 77-78, 81, 85, 95
Virtual Reality 7-8, 11, 24, 35, 42, 44, 62, 66, 73, 95, 103, 119, 122, 128, 163, 199, 241

www.ingramcontent.com/pod-product-compliance
Ingram Content Group UK Ltd.
Pitfield, Milton Keynes, MK11 3LW, UK
UKHW050418051026
15470UKWH00022B/1674